AF412740

RESEARCH PERSPECTIVES AND CASE STUDIES IN SYSTEM TEST AND DIAGNOSIS

FRONTIERS IN ELECTRONIC TESTING

Consulting Editor
Vishwani D. Agrawal

Books in the series:

Formal Equivalence Checking and Design Debugging
S.-Y. Huang, K.-T. Cheng
ISBN: 0-7923-8184-X
On-Line Testing for VLSI
M. Nicolaidis, Y. Zorian
ISBN: 0-7923-8132-7
Defect Oriented Testing for CMOS Analog and Digital Circuits
M. Sachdev
ISBN: 0-7923-8083-5
Reasoning in Boolean Networks: Logic Synthesis and Verification Using Testing Techniques
W. Kunz, D. Stoffel
ISBN: 0-7923-9921-8
Introduction to I_{DDQ} Testing
S. Chakravarty, P.J. Thadikaran
ISBN: 0-7923-9945-5
Multi-Chip Module Test Strategies
Y. Zorian
ISBN: 0-7923-9920-X
Testing and Testable Design of High-Density Random-Access Memories
P. Mazumder, K. Chakraborty
ISBN: 0-7923-9782-7
From Contamination to Defects, Faults and Yield Loss
J.B. Khare, W. Maly
ISBN: 0-7923-9714-2
Efficient Branch and Bound Search with Applications to Computer-Aided Design
X.Chen, M.L. Bushnell
ISBN: 0-7923-9673-1
Testability Concepts for Digital ICs: The Macro Test Approach
F.P.M. Beenker, R.G. Bennetts, A.P. Thijssen
ISBN: 0-7923-9658-8
Economics of Electronic Design, Manufacture and Test
M. Abadir, A.P. Ambler
ISBN: 0-7923-9471-2
I_{DDQ} Testing of VLSI Circuits
R. Gulati, C. Hawkins
ISBN: 0-7923-9315-5

RESEARCH PERSPECTIVES AND CASE STUDIES IN SYSTEM TEST AND DIAGNOSIS

by

John W. Sheppard
ARINC Incorporated

and

William R. Simpson
Institute for Defense Analyses

KLUWER ACADEMIC PUBLISHERS
Boston / Dordrecht / London

Distributors for North, Central and South America:
Kluwer Academic Publishers
101 Philip Drive
Assinippi Park
Norwell, Massachusetts 02061 USA
Telephone (781) 871-6600
Fax (781) 871-6528
E-Mail <kluwer@wkap.com>

Distributors for all other countries:
Kluwer Academic Publishers Group
Distribution Centre
Post Office Box 322
3300 AH Dordrecht, THE NETHERLANDS
Telephone 31 78 6392 392
Fax 31 78 6546 474
E-Mail <services@wkap.nl>

Electronic Services <http://www.wkap.nl>

Library of Congress Cataloging-in-Publication Data

A C.I.P. Catalogue record for this book is available
from the Library of Congress.

Preface

System level testing is becoming increasingly important. It is driven not by the march of technology, although that certainly has an impact, but by the incessant march of complexity. This growth in complexity is forcing us to renew our thinking on the processes and procedures that we apply to test and diagnosis of systems. In fact, the complexity defines the system itself which, for our purposes, is "any aggregation of related elements that together form an entity of sufficient complexity for which it is impractical to treat all of the elements at the lowest level of detail." System approaches embody the partitioning of problems into smaller inter-related subsystems that will be solved together. Thus, words like hierarchical, dependence, inference, model, and partitioning are frequent throughout this text. Each of the authors deals with the complexity issue in a similar fashion, but the real value in a collected work such as this is in the subtle differences that may lead to synthesized approaches that allow even more progress.

The works included in this volume are an outgrowth of the 2nd International Workshop on System Test and Diagnosis held in Alexandria, Virginia in April 1998. The first such workshop was held in Freiburg, Germany six years earlier. In the current workshop nearly 50 experts from around the world struggled over issues concerning the subject. These experts interacted in many ways throughout that week, including posters, formal presentations, tutorials, panels, and social events. In this volume, a select group of workshop participants was invited to provide a chapter that expanded their workshop presentations and incorporated their workshop interactions. While the selection was based solely on merit and did not include geographic distribution, these selections reflect the international flavor of the workshop. While we have attempted to present the work as one volume and requested some revision to the work, the content of the individual chapters was not edited significantly. Consequently, you will see different approaches to solving the same problems and occasional disagreement between authors as to definitions or the importance of factors. This is all part of the dialogue that advances the state-of-the-art.

To these workshop papers, we have added chapters on managing test information conflict, applying fuzzy logic in diagnosis, and developing standards for system test and diagnosis. These three topics are directly related to the application of the techniques covered in the other chapters. With these additions, we felt that a sharper picture of the system test and diagnosis discipline emerges.

As of this writing, the works collected in this volume represent the state-of-the-art in system test and diagnosis, and the authors are at the leading edge of that science. If you find a part of this dialogue interesting or useful to your work, we encourage direct correspondence with the authors and your participation in future workshops on the subject. We, the editors, are solely responsible for choice of the materials and will be delighted to discuss these or any related issues with interested parties. We can also provide information and materials on how to become a part of this dialogue on system test and diagnosis.

We would like to thank each of the authors for their contribution and for their cooperation in meeting tight publishing deadlines and our unreasonable demands. We also thank members of the community, especially the workshop participants. Each helped to formulate the ideas and approaches presented in this volume.

J. W. Sheppard
W. R. Simpson
Annapolis.

List of Contributors

Anthony P. Ambler is currently with the Department of Electrical and Computer Engineering at the University of Texas at Austin. Prior to that he held the Chair in Test Technology at Brunel University, UK. He has worked for several years in the area of test process optimization for chips, boards, systems and field test. He is a Fellow of the IEEE. Dr. Ambler can be reached at ambler@exe.utexas.edu.

Darrell Bartz is a Senior Project Engineer assigned to the Boeing Phantom Works–Vehicle Systems Technology. Mr. Bartz has been working Internal Research and Development projects for 6 years, most recently he has been responsible for development and implementation of Boeing's Knowledge Discovery and Data Mining efforts in St. Louis. Mr. Bartz received a B.S. in Electronic Engineering Technology from Lake Superior State University and an M.S. in Management Information Systems from the University of Missouri–St. Louis. Since joining McDonnell Douglas 14 years ago he has been involved in all phases of aircraft support and diagnostics development. He established and leads the Diagnostic Engineering Team for the F/A-18E/F Program. Mr. Bartz can be reached at darrell.o.bartz@boeing.com.

Moshe Ben-Bassat has been president of IET Intelligent Electronics and Professor of Computers and Information Systems at Tel-Aviv University since 1983. During 1975-1983 he was at Los Angeles, California where he was Program Director of MEDAS (an Expert System for Medical Diagnosis) at the University of Southern California and a Consultant for a wide variety of projects supported by DARPA and others. At IET, he initiated, and now guides, the development of the TechMate and W-6 products. Dr. Ben-Bassat can be reached at Moshe@notes.iet.co.il.

Timothy M. Bearse received a BS degree in Electrical Engineering from the Southeastern Massachusetts University in 1974, and the MS degree in Electrical Engineering from the University of Rhode Island in 1993. He has worked for the Naval Undersea Warfare Center since 1974 performing work in the area of test systems engineering and technology. He has performed hardware and software engineering, fleet support, and systems engineering efforts in support of missile test systems. He presently serves as project engineer for the Integrated Diagnostic Support System (IDSS) program where he has responsibility for integrated diagnostics technology and applications of the IDSS tool set. Mr. Bearse can be reached at bearsetm@code831.npt.nuwc.navy.mil.

Israel Beniaminy has been Technology Director at IET since 1994 and is responsible for specifying the development of new IET products and guiding the adaptation of IET products to new environments. Israel Beniaminy graduated from Bar-Ilan University with a B.Sc. in Physics and Computer Science in 1981, and joined IET in 1986. He has extensive experience in advanced software technologies and algorithms, and in applying them in the test and maintenance world. Mr. Beniaminy can be reached at IsraelB@notes.iet.co.il.

Anton Biasizzo received the B.Sc. and M.Sc. degrees in electrical engineering from the University of Ljubljana, Slovenia, in 1991, and 1995, respectively. He is currently working toward his Ph.D. degree in the field of sequential system diagnosis. He is a research assistant at Jožef Stefan Institute since 1991. His research interests are efficient algorithms for sequential diagnosis based on AND/OR graphs, constraint logic programming in model based diagnosis and automatic diagnostic test pattern generation. Mr. Biasizzo can be reached at anton.biasizzo@ijs.si.

Wai Chan is a statistician in the System Business Unit at Digital Equipment Corporation. Previously he taught statistics at the Ohio State University. He received his Ph.D. in statistics at Florida State University under the direction of Frank Proschan and Jayaram Sethuraman. Dr. Chan can be reached at chan@usctrl.enet.dec.com.

Harry Dill is founder and President of Deep Creek Technologies, Inc. He holds a Bachelor's Degree in Electrical Engineering from Drexel University, a Masters of Engineering Science Degree from Loyola College and has 25 years of experience–15 of those years in diagnostic testing. Mr. Dill was a principal contributer in the development of Intusoft's Test Designer, a SPICE simulation based test synthesis tool and has performed numerous test synthesis tasks using that tool. Related assignments include development of diagnostic strategies for satellite communication systems. Mr. Dill can be reached at TestDesigner@compuserve.com.

Des Farren is manager of software test and support operations in the GSM group at Motorola, Cork, Ireland. Previously he was the test engineering manager at Digital Equipment, Ayr, Scotland, and, still earlier, Galway, Ireland. While at Digital, he carried out the work presented in this article and completed a Ph.D. at Brunel University, Uxbridge, UK. Farren is a member of the IEEE and the Computer Society. Dr. Farren can be reached at farrend@cork.cgi.mot.com.

Jeffery Holland is currently working as Program Manager for an advanced mission processing research project at the Boeing Company. His previous work included the development and demonstration of a prototype onboard diagnostic system. This system used neural network and fuzzy logic software technology to monitor multiple avionics systems and provide an intelligent integrated diagnosis and some degree of prognostics. He also has experience designing diagnostic equipment for military maintenance personnel for F-15E and AV-8B aircraft avionics. He was recently involved in the development of avionics software and support libraries for commercial grade single-board computers. He holds a masters degree in electrical engineering from Washington University. Mr. Holland can be reached at jholland@boeing.com.

David Joseph is Vice President of Sales, North America TechMate Division, and joined IET in 1993. David graduated from Ben-Gurion University with B.Sc. in Computers & Electric Engineering in 1984. David served in the Israeli air force for 12 years in the capacities of training, maintenance, and service of wide verity of radar and avionics equipment. David has extensive experience in Image Processing Systems and Computer Driven products, which implement Artificial Intelligence algorithms for testing and service. Mr. Joseph can be reached at davidj@ietusa.com.

Kirby Keller is a Senior Principal Technical Specialist for the Boeing Company. He is the principal investigator of several research and development projects that are applying data mining, neural network and fuzzy logic technologies to diagnostics/prognostics and health management. He is the technical lead on the Integration Standards and Performance Metrics for the Next Generation Aircraft program sponsored by the Air Force Research Laboratories at Wright-Patterson which is developing a software architecture for information fusion. He was a system engineer on the U. S. Army RotorCraft Pilot's Associate program where his duties included developing and executing an approach to requirements analysis, software architecture definition and system engineering. He holds a Ph.D. in Applied Mathematics from Iowa State University and is the author of over 30 technical papers, conference presentations, and technical reports. He is a member of AIAA, IEEE and ACM. Dr. Keller can be reached at kirby.j.keller@boeing.com.

Michael L. Lynch has been an electronics engineer at the Naval Undersea Warfare Center in Newport, RI since 1983. He works on projects related to test technology and integrated diagnostics and is currently involved with the commercialization of the Integrated Diagnostics Support System (IDSS) tool set. He earned a B.S. in electrical engineering at University of Massachusetts at Dartmouth in 1983 and a M.S. in electrical and computer engineering at the University of Massachusetts at Amherst in 1990. Mr. Lynch can be reached at lynchml@code831.npt.nuwc.navy.mil.

Franc Novak gained the B.Sc., M.Sc. and Ph.D. degrees in electrical engineering from the University in Ljubljana, Slovenia, in 1975, 1977, and 1988, respectively. Since 1975 he has been with the Jožef Stefan Institute, where he is currently head of Computer Systems Department. He has also been an associated professor at the University of Ljubljana. His research interests include electronic design, test and diagnosis. His most recent assignment has been on design for testability of analog circuits. He has published a number of articles and papers in the field of electronic test and diagnosis. He was a member of Technical Programme Comitte of European Design and Test Conferences 1995-97, and DATE'98 Conference. He is a member of IEEE and a member of NORMATE (Network of Researchers in Mixed-signal and Analog Testing). Dr. Novak can be reached at Frank.Novak@ijs.si

John W. Sheppard is a Staff Principal Analyst at ARINC Incorporated. He holds a BS in Computer Science from Southern Methodist University and an MS and Ph.D. in Computer Science from Johns Hopkins University. He has twelve years experience performing research in artificial intelligence applied to test and diagnosis and has published over 80 technical articles in the test area. Also, he is a co-author with William R. Simpson of the book, *System Test and Diagnosis*. Currently, Dr. Sheppard is the Computer Society Liaison to the IEEE Standards Coordinating Committee 20, co-chair of the IEEE Computer Society System Test Technical Activity Committee, co-chair of the AI-ESTATE subcommittee of the IEEE Standards Coordinating Committee 20, and is US delegate to IEC/TC93 and secretary of IEC/TC93/WG7 (System Test). Dr. Sheppard can be reached at jsheppar@arinc.com.

William R. Simpson is a professional staff member at the Institute for Defense Analyses where he is involved in defining software architectures and the commercial standards associated with automatic test systems. Before joining IDA he was a research fellow in the Advanced Research and Development Group at ARINC Research Corp. where he was involved in testability and fault diagnosis and helped to develop the System Testability and Maintenance Program (STAMP) and the POrtable Interactive TroubleshootER (POINTER). Dr. Simpson is author of over 100 publications and recently a co-author of the textbook *System Test and Diagnosis*. He chairs the IEEE Standards Coordinating Committee 20, co-chairs the IEEE Computer Society System Test Technical Activity

Committee, and is US delegate to IEC/TC93 and co-convener of IEC/TC93/WG7 (System Test). He holds a BS degree in Aerospace Engineering from the Virginia Polytechnic Institute and State University, MS and Ph.D. degrees in Aerospace Engineering from Ohio State University, an MSA in Engineering Administration from The George Washington University, and is a graduate of the US Naval Test Pilot School. Dr. Simpson can be reached at rsimpson@ida.org.

Kevin Swearingen is a Project Engineer in the Boeing Company's Phantom Works. His current responsibilities are focused on applications of data mining technologies to improve aircraft prognostics and health management capabilities. Previously, he implemented neural network, fuzzy logic, and expert system solutions on several diagnostic development projects. He was involved in developing support for the T-45TS US Navy jet trainer program and avionics support for the AV-8B US Marine Harrier jump jet. He was awarded his MSEE in 1997 and BSEE in 1988 from the University of Missouri - Rolla. Mr. Swearingen can be reached at kevin.j.swearingen@boeing.com.

Jack Taylor has been involved involved with testing of Flight systems on the prototype and pre-production Concorde SST. Later, he moved on to design, test and manufacture of missile systems. Now semi-retired he works part time for APSYS Ltd. Jack has been a member of IEEE C/ATLAS 716 and AI-ESTATE subcommmittees of IEEE SCC20 since 1990. Mr. Taylor can be reached at JackTaylor@compuserve.com.

Alenka Žužek received the B.Sc. degree in electrical engineering from the University of Ljubljana, Slovenia, in 1991, and the M.Sc. degree in computer science from the same university in 1997. Currently, she is working toward the Ph.D. degree in the field of CAD for the integrated circuits. She is a research assistant at Jožef Stefan Institute since 1993. Her research interests include AND/OR graph search algorithms for sequential diagnosis, and the use of decision diagrams for problems in CAD. Ms. Žužek can be reached at Alenka.Zuzek@ijs.si.

Table of Contents

Chapter 1

Diagnostic Inaccuracies: Approaches to Mitigate

William R. Simpson
Institute for Defense Analyses

Keywords: Outcome-based testing, uncertainty, test limits, diagnosis, testability, confidence factors, cannot duplicate, false alarm, false assurance.

Abstract: Inaccurate diagnostics cause wasted man-hours, increased inventory and general levels of frustration about maintenance. Often it seems that the more costly equipment is the most flagrant offender. After we have fixed the blame and fixed the problem for technician error, bad testability, and improper documentation, there will still be a residual level of inaccuracy based upon the outcome-based testing that we currently do. This chapter will explicitly show the limitations imposed by measurement theory and provide useful solutions for attempting to limit the inaccuracies from measurement inaccuracies and outcome representations. While laboratory scientists have been aware and apply these techniques, test engineers developing test for field maintenance seldom consider these factors. These inaccuracies lead to occurrences of false alarm, false assurance and field cannot duplicate events and Re-Test OK. Understanding the base causes and how to approach resolution will assist in the development of diagnostic and repair strategies.

1.   BACKGROUND

The current approach to testing intertwines the test procedures, the sequence of tests, and the diagnostic outcomes. Typically testing involves a series of stimulus, measurement, and compare events that categorize the answer as either *good* or *bad*, the former meaning that measured conditions are within nominal circumstances and the latter meaning that they are not within nominal circumstance. The latter is normally taken to mean that a fault has been detected. Diagnoses occurs as follows:

- Retest OK when the collection of tests is nominal
- Infer Anomaly is a member of [list] for anomalous outcome
- Anomaly is intersection of all [lists] from anomalous outcomes, or the [list] associated with an anomalous outcome.

Such black and white interpretation of outcomes is a major contributor to inaccurate diagnosis in many instances. Addressing multiple outcomes such as fail-high, fail-low, etc. may do little if anything to alleviate the problem. The key lies in measurement science and its application to the test problem.

2. MEASUREMENT SCIENCE AND "ACCURACY"

Under ideal conditions, with "accurate" measurement devices and full information about the system under test, diagnostic errors will still occur. These are mathematical residues that cannot be eliminated completely. It is important to understand that some complicated elements of diagnosis involve chasing these residues.

A prominent issue in testing is the impact of precision and accuracy of test resources on the certainty in the resulting test outcome. Formalizing the impact of precision and accuracy on test confidence comes from work in measurement theory. Given the need to formalize these factors, we can consider each of them in terms of probability distributions.

The precision of a resource characterizes the amount of "scatter" one can expected when repeated measurements are made of the same signal. Typically, precision is applied to the measurement process or resource rather than the measurement itself. Thus the precision of a resource can be characterized by considering a probability distribution of measurements. A precise resource would yield a narrow distribution of measured values (i.e., a low variance or standard deviation), while an imprecise resource would yield a wide distribution of measured values (i.e., a high variance or standard deviation).

On the other hand, the accuracy of a measurement corresponds to the amount of "error" in the measurement. In other words, accuracy indicates a level of deviation from some *reference value*. As with precision, accuracy can be modeled as a probability distribution. Typically, accuracy is applied to the measured value rather than the process the instrument used in taking the measurement. Frequently, it is determined through taking several independent measurements and taking the mean deviation from some reference value.

Inaccuracy can result from two sources—random error and systematic error. Systematic error is also called bias. Many consider accuracy only in terms of random error (referred to as an unbiased estimate of error) in which bias is ignored or calibrated out. Others consider total error in which the bias is included.

3. SETTING TOLERANCES ON UUT BEHAVIOR

To determine the appropriate tolerances for a particular test, and thereby determine relevant test outcomes, one must consider the precision of the required resources. The precision will be used to determine the associated bias of the instrument. To determine the precision, typically, several independent measurements are taken for that instrument under known conditions and the resulting distribution determined. Then the width of the distribution is determined by using, for example, variance. Since the sources of error for measurements come from numerous sources, it is often assumed that the central-limit theory applies and normal distributions apply. For this work, we will assume normally distributed while generalizing the conclusions beyond the normal distribution.

Next, measurement error is considered by examining the distribution characterizing accuracy. Given the system to be tested and a measurement to be made, a distribution of "nominal values" for that measurement can be determined. From this distribution, in the simplest case, Pass/Fail criteria are established based on the probability of a measurement occurring within some set of defined limits applied to that measurement (Figure 1).

It is generally the case that the Pass/Fail criteria are determined by considering expected values for a fault-free unit. "Significant" deviation from these expected values result in the fail outcome for that test. The limits define what is meant by "significant." During testing, a measurement value is typically mapped into a discrete outcome determined based upon which side of these limits the measurement falls.

Next we add a measurement to the UUT distribution as shown in Figure 2. The figure shows two instruments. The first (shown on the left-hand side of figure 2) is tightly distributed compared to the UUT and provides a very low probability of assigning the wrong outcome. Recall that the probability of assigning a wrong outcome is related to the area under the curve. This is given as

$$P(x_1 < M \vee M > x_2) = \int_{-\infty}^{x_1} f(x)dx + \int_{x_2}^{\infty} f(x)dx$$

where M is the measurement and P represents the probability of the measurement being in any interval that will provide a misidentification.

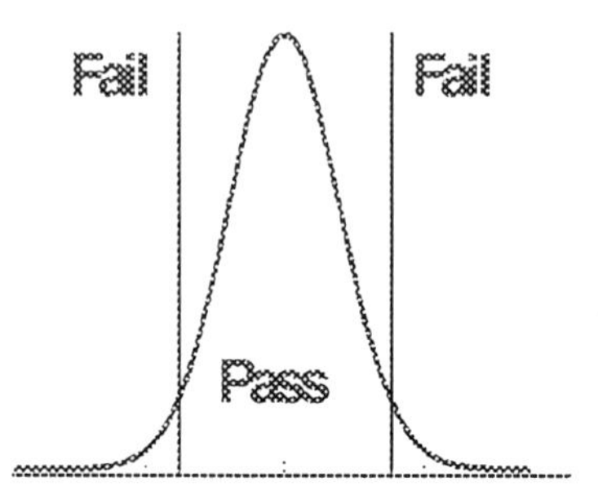

Test
- Stimulate the UUT
- Perform Measurement
- Compare to Nominal
- Assign Outcome

Diagnose
- RTOK when the collection of tests is nominal
- Infer Anomaly is a member of [list] for anomalous outcome
- Anomaly is intersections of all [lists] from anomalous outcomes, or the [list] associated with an anomalous outcome.

Figure 1. Setting Pass/Fail Criteria for a Unit Under Test

The tighter the measurement distribution is when compared to the UUT distribution, the smaller the probability of error. This becomes very low when the ratio is about 10. In fact this is where the rule of thumb for measurement needing ten times the accuracy of the item being measured. The second (on the right-hand side of figure 2) shows an instrument with about the same accuracy as the UUT. Here we see that there is a finite and not small possibility that the outcome can be designated wrongly. In fact, we can actually compute the probability of this measurement being wrong. From our basic probability background we know this as the area under the measurement curve that occurs to the left of the threshold value. Now, we can only do this because the problem is fully explicit. That is, we know distributions, thresholds and the "actual value". The trick of course, is to measure ten times as accurately as the UUT value, but this may not be possible when we are pushing the state-of-the-art. In fact, when pushing the state-of-the-art the measurement capability may be equal to the UUT capability. It can be seen now why the more complex and expensive devices have a bigger problem with accurate diagnosis. It is these units that are pushing the state-of-the-art.

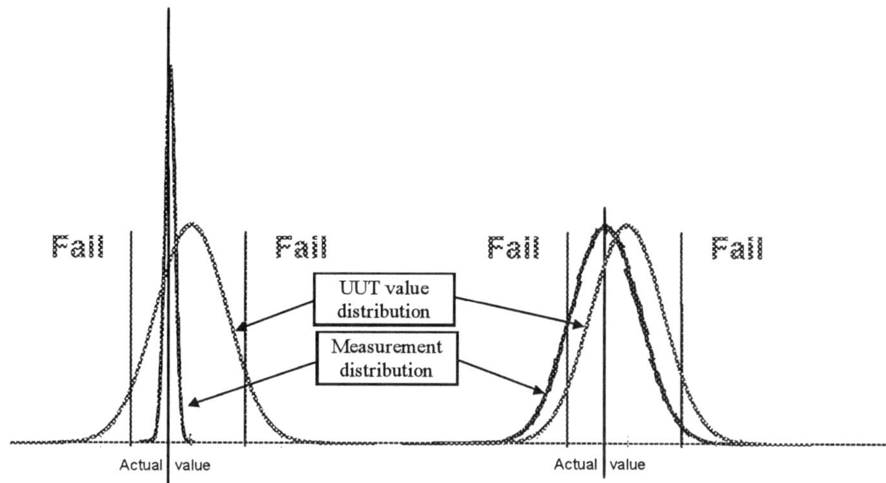

Figure 2. Measurement distribution superimposed on UUT distribution

Once the measured value is mapped into the discrete outcome, interpreting the test results becomes problematic. Specifically, when considering a single measurement, the probability of that measurement being within tolerance will be $P(x)$, and the probability of it being out of tolerance will be $1 - P(x)$. (Actually, these are conditional probabilities since the actual probability depends on whether or not the unit has failed.)

The problem of misidentified outcomes becomes particularly acute when the measurement is near the threshold itself. Here even normal rules of thumb do not apply, as shown in Figure 3. We have provided two samples from the same instrument one providing a high probability of a good outcome, and the other a high probability of a fail outcome. The more pointed distribution is the measurement, and the peak of the distribution could be taken as the "actual" measurement. Both may actually be at the same point in "reality", but at different points in the measurement distribution. In the left side of figure 3, the peak of the measurement distribution is inside the pass area, but the measurement itself can be anywhere in the first distribution giving it a more than 50% probability of passing. On the right, the peak is outside the pass area and it has more than a 50% probability of failing. In an actual event, the peak value "measure" in each instance may be for the same identical situation. The conclusion is that values near thresholds are always suspect.

As discussed earlier, most test tolerances are determined given the distribution of values for a fault-free unit (i.e., assuming the test should pass). Different failure modes, however, exhibit different types of behavior and also generate distributions of measurement values. Understanding how

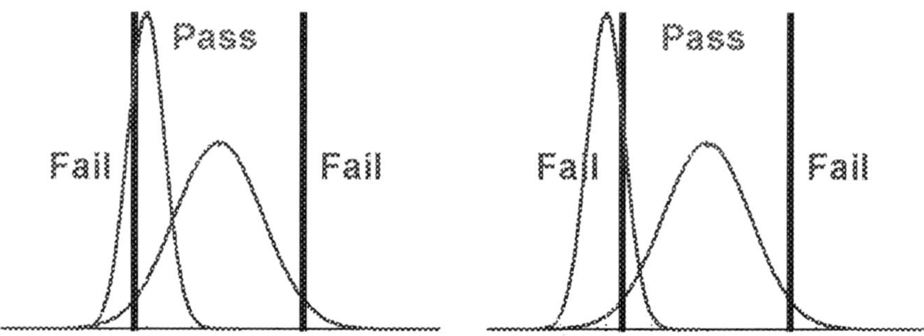

Figure 3. Errors Arising Near Thresholds

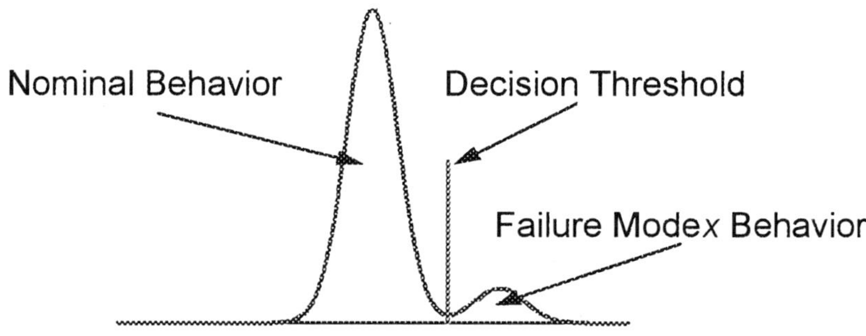

Figure 4. Behavior Distribution

these failure modes are manifest can be used to refine the definitions of tests and the tolerances on those tests.

If we consider a combined distribution where the population includes fault-free as well as faulty units, we may begin to see multi-modal distributions. These distributions may be constructed with representative samples of faulty and fault-free units, based on failure rate information for each of the failure modes. Under these conditions, the various modes of the distribution with correspond to the various states (or "modes") of the unit (Figure 4). Identifying these behaviors are also complicated by measurement inaccuracies.

The above scenario only works fairly well when considering isolated elements in a unit and directly measuring the presence or absence of unit characteristics. In order to work at all, they require the defined failure modes to be distinct from normal characteristics and each other. When thresholds begin to overlap or even become adjacent, the problem of measurement error again has a major impact. Failure mode identification is also subject to the effects of measurement error when actual value thresholds are approached. Unfortunately, the process of capturing and modeling unit behavior (either

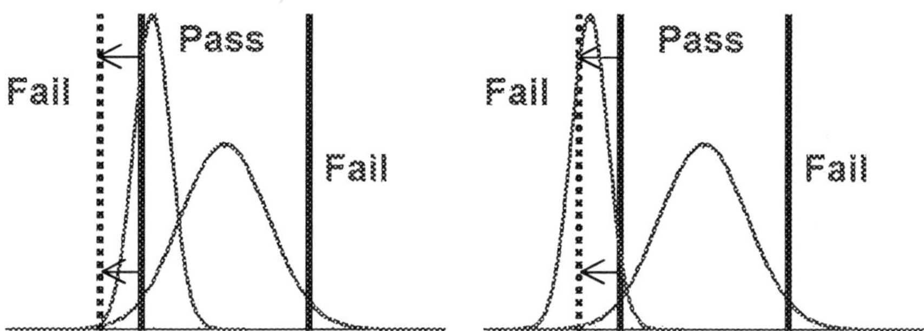

Figure 5. Expanded Thresholds

faulty or fault-free) becomes very complex within the context of a system. Expected behavior of a component evaluated outside of a system (e.g., with direct probing) is likely to be very different from the same component evaluated within the system (e.g., from the gold fingers).

Variations in behavior arise from interactions between components in the system, thus (significantly) perturbing the expected values when measurements are taken (except in very simplistic cases). The end result is complication in determining tolerances on tests for diagnosis. Thus the ability to re-use tests is largely limited to re-use of test methods only. Tolerances, confidences, and inferences must be re-engineered for each type of system.

4.　　ATTEMPTS AT HANDLING THE PROBLEM

The normal response to a test that fails too many good units is to open up the tolerances. While the immediate effect is to slow down the rejection of fault-free units, it is apparent that the acceptance of faulty units is increased. Figure 5 illustrates the fallacy of this approach. While setting the threshold out, we certainly reduce the chances of rejecting a measurement on the good side of the threshold. Again, we will use the convention of identifying the more pointed distribution as the measurement, and the peak to be the "actual" value. We see that on the left figure opening up the tolerance band reduces the probability of rejecting a unit close to the original threshold to near zero, but the right side of the figure shows a sizeable probability of accepting a unit some distance from the original threshold. In general, the tightening of tolerances will decrease the probability of accepting a faulty unit, while opening tolerances will decrease the probability of rejecting a good unit. The trick is to balance these factors to get the "best" maintenance set. There is no clear or consensus definition of "best". Nor should there be such a definition. The needs are extremely dependent on the context of the problem. For example, safety of flight may dictate tight tolerances, while convenience instruments, or sufficiently backed up capabilities might adhere

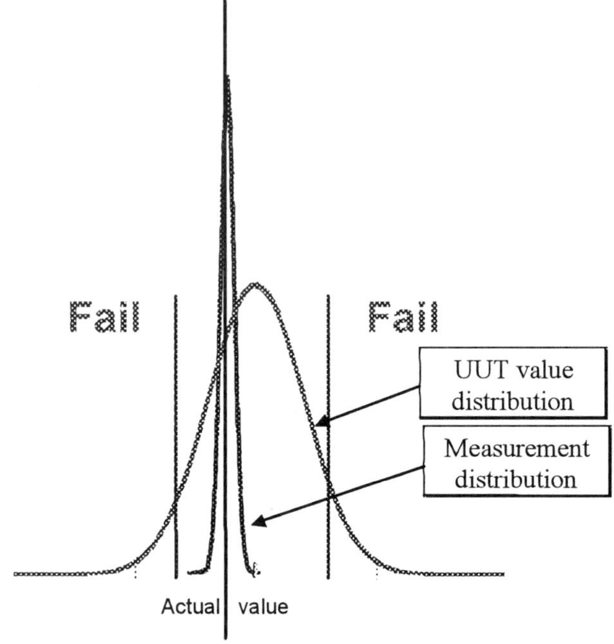

Figure 6. Case 1 High degree of Accuracy

to looser tolerances to reduce maintenance expense. The most important lesson here is that dealing directly with outcome based testing is subject to errors and we must either accommodate them or seek ways to minimize them.

We can examine the outcome base testing as a series of cases:

1. High degree of accuracy in measurement
2. Equivalent degree of accuracy between measurement and UUT precision.
3. Near a threshold

Case 1. High Degree of Accuracy is typified by a relatively pointed distribution as shown in Figure 6. We have taken the case where the measurement is not near a threshold, since that is specifically covered in case 3. Figure 6 shows a low probability of error, in fact in the figure the integrals are zero. This has led to the famous rule-of-thumb to measure ten times more accurately than the number you are measuring. Fair statement for the lab, but difficult at best with field diagnostics.

Case 2. Equivalent Accuracy is typified by similar distributions as shown in Figure 7. We have again taken the case where the measurement is not near a threshold, since that is specifically covered in case 3. Figure 7 shows that the measurement has a significant probability of being in the wrong domain for

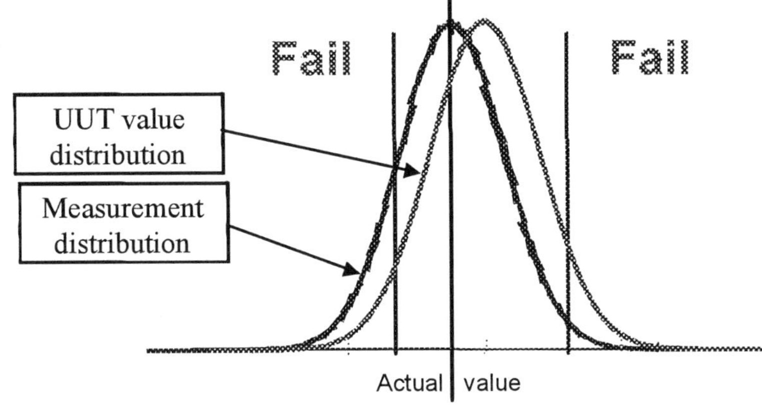

Figure 7. Case 2 Equivalent degree of Accuracy

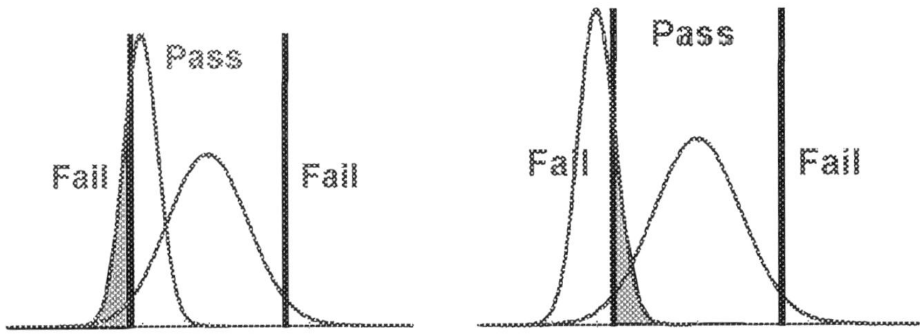

Figure 8. Case 3 Near a Threshold

identifying the test outcome correctly and has a significant probability of assigning the wrong outcome.

Case 3. Near a Threshold is shown in Figure 8, for both a nominally good system (first half of figure), and a nominally bad system (second half of figure. Figure 8 shows that the measurement has a significant probability of being in the wrong domain for identifying the test outcome correctly, even when accuracy is high, and has a significant probability of assigning the wrong outcome.

Case Summary–While the first case certainly presents no problem in misidentification, the second and third cases do. The second case regularly occurs when we are pushing the state-of-the-art and no instrument approaching the ten-to-one threshold is available. It may also occur when we are pushing the economic values of a project and someone has made a trade-off concerning test equipment costs. The last case exists in all testing, and moving thresholds to accommodate just makes the problem worse. The

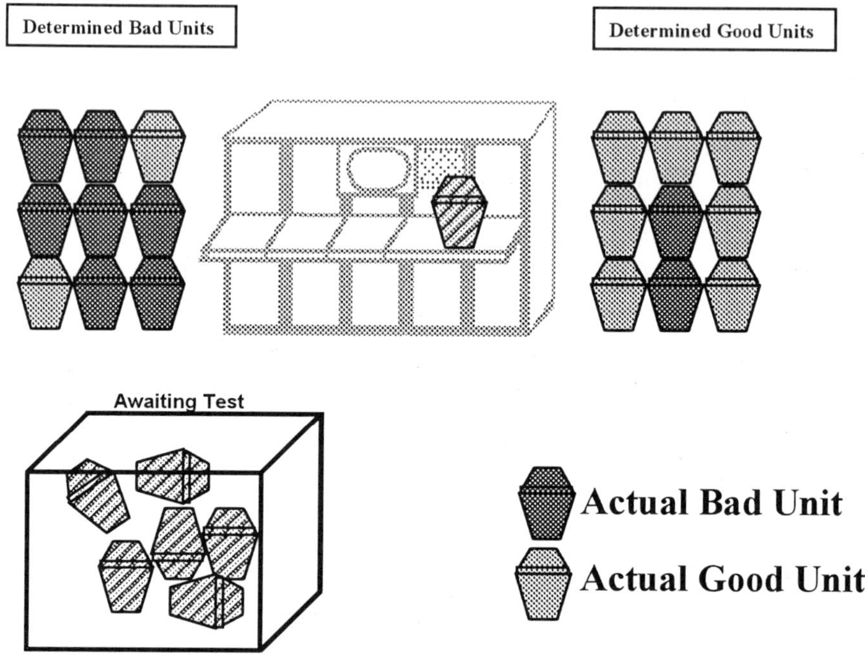

Figure 9. Outcome misidentification consequences

bottom line is, that we should buy the best test equipment we can afford, and always be concerned about the accuracy of test outcomes in outcome-based testing. The consequence of ignoring misidentified test outcomes is shown in figure 9. The cannot duplicate or Re-Test OK events will be represented by the lighter boxes on the left. These are manifest and becoming a growing concern as electronic complexity increases. Just as serious, but less obvious in the field are the dark boxes on the right which represent a false assurance that systems may operate up to expectations. These will manifest themselves only through aborted and failed operations.

5. MOVING TEST TOLERANCES REVISITED

The above discussion provides a background around which to discuss the manipulation of test tolerance to achieve certain objectives of the particular test element. We will discuss three common approaches to the problem:

- Repeat testing
- Opening Tolerances
- Closing Tolerances

Repeat Testing–When the test is being conducted as a Built-In-Test (BIT), misidentification of a test on the faulty side is termed false alarm. If the misidentification of the fault occurs too frequently, the indication is ignored completely, and as such, BIT losses its effectiveness. One common response is to repeat the test a number of times offering it the option of not providing a wrong indication. This is sometimes called repeat polling. This actually works on two problems characteristic of BIT systems. The first is that BIT may detect transient conditions indigenous to the operation of the system. The transient condition may be real, or an artifact of the BIT system. The repeat polling allows the transient to subside. In fact, this leads to an approach to testing that says something like only "indicate an anomalous condition if you have three repeat indications within x milliseconds". This, of course assumes that the three measurements are independent and the "good" indication will override two "bad" indications. Neither assumption is valid, and the approach has even more disastrous effects when we consider that the outcome-based testing carries inaccuracies that are measurement derived. The result is, that repeat polling leads to fewer false alarms, but higher missed detection rates and greater false assurance. This in turn may lead to lost operations, or allow a faulty product to get into the hands of a user.

Opening Tolerances–A second approach to mitigating high reject rates is to open tolerances. Figure 10 illustrates what is happening from the standpoint of the distributions and the outcome based approach.

Opening tolerance the amount shown in Figure 10a almost guarantees a "good" unit will pass. We have again taken the assumption that the "true" value is the peak of the measurement curve. The amount needed to open up is a function of the relationship between the measurement accuracy and the UUT accuracy. Unfortunately, this will increase dramatically the probability of a "bad" unit passing as shown in figure 10b. The resulting mix is illustrated in figure 11.

The more interesting result of course, is that this looks like an improvement. Fewer units are rejected. In a manufacturing environment "apparent" yield goes up and in the field, cannot duplicate events are reduced. For BIT, we have a reduction in "apparent" false alarms. It is a bit of false economy because the inaccuracy has been pushed to the other type error and increased overall. False assurance will lead to higher instances of customer complaints and in operational systems we may have lost operations and excessive down-times for completion of "special" diagnostic investigations.

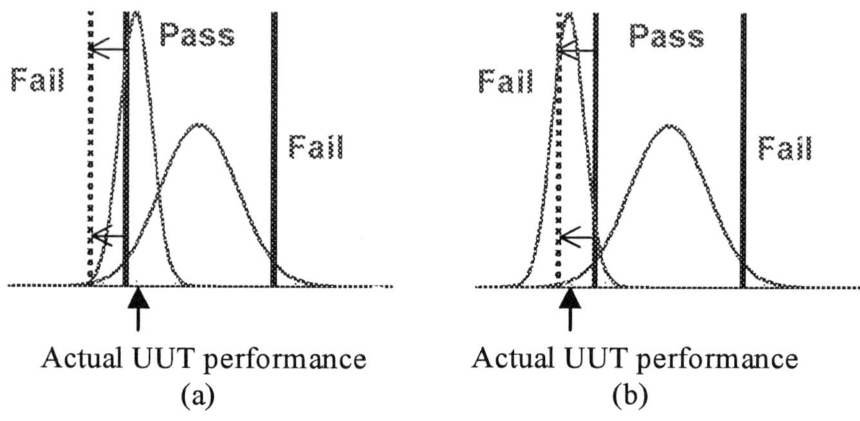

Figure 10. Opening Tolerances

<u>Closing Tolerances</u>–The final approach to be discussed is to close tolerances. This is usually undertaken as a response to high "leakage" (allowing bad units out of the factory to a customer), or large losses in field performance. Figure 12 illustrates what is happening from the standpoint of the distributions and the outcome-based approach.

Closing tolerance the amount shown in figure 12a almost guarantees a "bad" unit will fail. We have again taken the assumption that the "true" value is the peak of the measurement curve. The amount needed to close up is a function of the relationship between the measurement accuracy and the UUT accuracy. Unfortunately, this will increase dramatically the probability of a "good" unit failing as shown in figure 12b. The resulting mix is illustrated in figure 13.

The more interesting result here, is that this looks like an increase in the reject rate leading to higher cannot duplicate events. Fewer units are accepted. In a manufacturing environment "apparent" yield goes down and in the field, cannot duplicate events are increased. For BIT, we will see higher "apparent" false alarms. Despite all the drawbacks, there are good sound engineering reasons to close down tolerances. One excellent reason is for safety critical systems where we cannot afford to be wrong. The insurance we pay is increased false alarms and cannot duplicate events. A second is customer satisfaction, where we are willing to absorb additional costs to keep the customer happy. Finally, in providing strong warranties and performing field repair of customer systems, we may actually reduce overall costs by tightening tolerances.

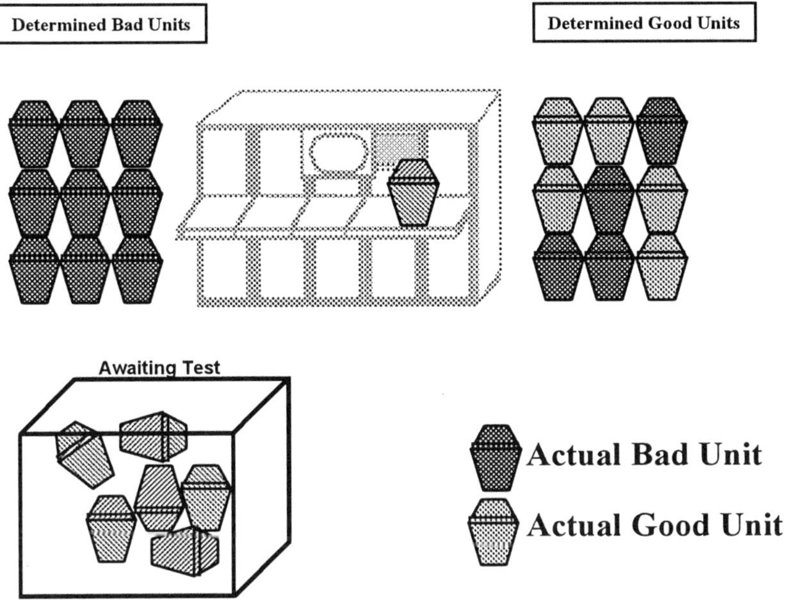

Figure 11. Opening Tolerances – Test Results

6. SIMULATION, UNCERTAINTY, AND LEARNING

Additional techniques that may help with diagnostic errors can be characterized as simulation, uncertainty, and learning approaches.

<u>Simulation</u>–Given the complexity associated with manually analyzing a system to determine performance and test tolerances, fault simulation is frequently used as an approach to analyzing effects of failure modes within a system. Unfortunately, using simulation as the basis for generating tests and diagnostics is very expensive. Incorporating models for error sources or instrument precision make the simulations even more accurate, and even more expensive.

One approach to reducing cost is to model the test error (thereby determining test outcome base confidence) and test-to-fault confidence separate from the simulation. The former restores the simulation to the case where one assumes perfect instrumentation, and the latter reduces the scope of the simulation by restricting the space of possible faults simulated. Unfortunately, no matter how robust the models or the simulation, situations will always occur in which results are in error. The best way to deal with these situations is to have a diagnostic capability employing some form of reasoning under uncertainty.

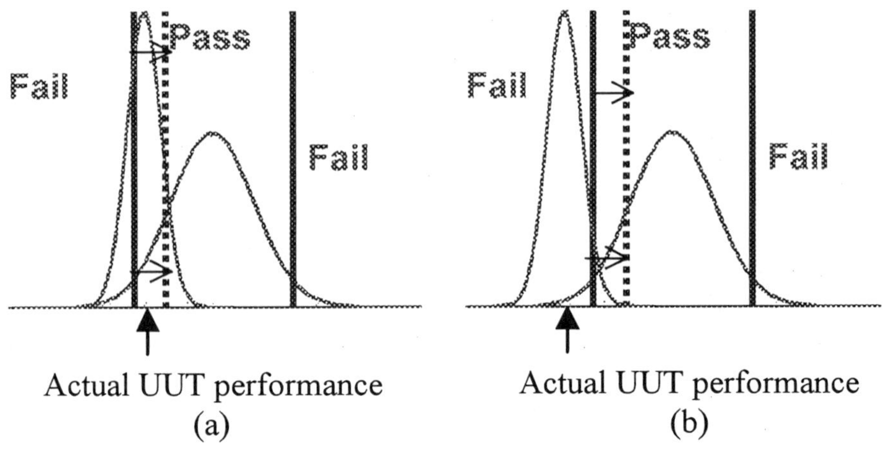

Figure 12. Closing Tolerances

Uncertainty–The uncertainty may be handled in a number of ways (Simpson and Sheppard, 1994) including the assignment of a base confidence. In determining the base confidence (i.e., either the expected confidence or the maximum confidence) in the outcome, we want to consider $P(x$=in-tolerance$|c$=good) and $P(x$=out-of-tolerance$|c$=bad). As shown in Figures 5 and 9, problems occur when a measurement for a good unit is out of tolerance (referred to as a "Type I error") or a measurement for a bad unit is in tolerance (referred to as a "Type II error"). A number of tools exist to handle the uncertainty calculations and there are even IEEE standards (IEEE, 1995; IEEE, 1997) providing representation and services for diagnostic models including uncertainty measures.

An estimate of the probabilities involved may be made if we make a number of simplifying assumptions such as measurent and accuracy being identically distributed and independent. While these assumptions are limiting, they may be used for a first cut on probabilities using on UUT and instrument specification data applied to the individual measurements. The technique is shown in figure 14, where the distribution is derived from the instrument specification.

Machine Learning–Finally, there is a great body of literature concerning machine learning techniques where differences between individual UUTs and instruments may be learned and accounted for after some training or field experience. It is almost never a good exercise to open tolerances, although it may solve a short term political problem (excessive false alarms, low yields, high cannot duplicates). The one exception is in high cost convenience items where loss of the subsystem will have little impact on field performance and we wish to be certain that the item is faulty before we undertake repair.

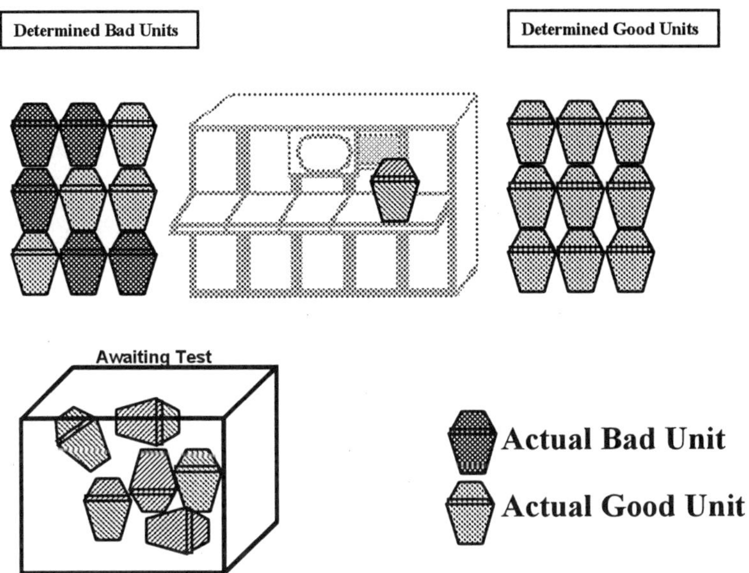

Figure 13. Closing Tolerances – Test Results

7. SUMMARY

We have shown that under the most optimal conditions, diagnostic errors will still be present in system level test. The problem is amplified when UUT complexity or pushes on the state-of-the-art. Conventional approaches to the common manifestations only provide temporary solutions with longer term costs. As a result, the outcome based testing should treat all tests as suspect or uncertain. Manually determining test tolerances is inaccurate and expensive. Fault simulation is frequently used as an approach to analyzing effects of failure modes within a system. Unfortunately, using simulation as the basis for generating tests and diagnostics is very expensive. Incorporating models for error sources or instrument precision makes the simulations even more accurate, and even more expensive.

Other methods include multiple polling or repeated instances of tests to improve outcome accuracy, and of course, you can always buy more accurate instruments if they are available. This, of course, does not eliminate the problem, but may minimize it.

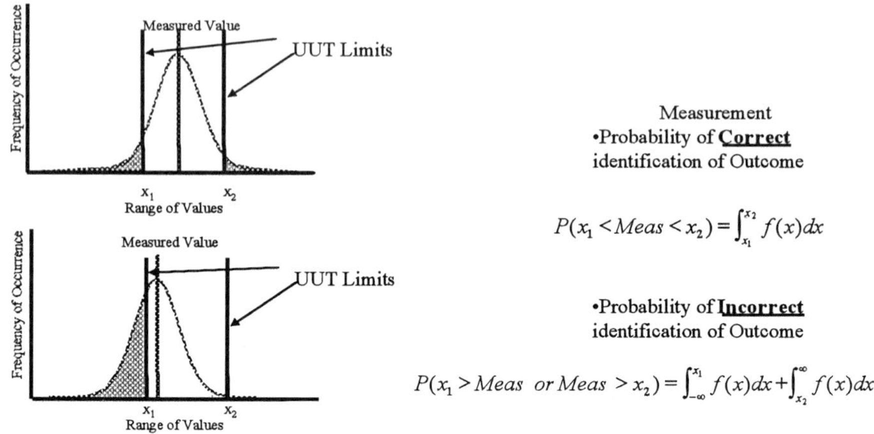

$$P(x_1 < Meas < x_2) = \int_{x_1}^{x_2} f(x)dx$$

$$P(x_1 > Meas \ or \ Meas > x_2) = \int_{-\infty}^{x_1} f(x)dx + \int_{x_2}^{\infty} f(x)dx$$

Figure 14. Estimating Probabilities Associated with Measurement

8.   REFERENCES

IEEE Std 1232-1995. *Trial Use Standard for Artificial Intelligence and Expert System Tie to Automatic Test Equipment (AI-ESTATE): Overview and Architecture.*

IEEE Std 1232.1-1997. *Trial Use Standard for Artificial Intelligence Exchange and Service Tie to All Test Environments (AI-ESTATE): Data and Knowledge Specification.*

IEEE P1232.2. (1997, August). *Draft Trial Use Standard for Artificial Intelligence Exchange and Service Tie to All Test Environments (AI-ESTATE): Service Specification,* Draft 4.0.

MIL-STD-2165. (1985, January). *Testability Program for Electronic Systems and Equipments.*

Keiner, W. 1990. "A Navy Approach to Integrated Diagnostics," *Proceedings of AUTOTESTCON '90,* IEEE Press, pp. 72-76

Simpson, W. R. and J. W. Sheppard. 1994. *System Test and Diagnosis,* Norwell, MA: Kluwer Academic Publishers.

Chapter 2

Pass/Fail Limits—The Key to Effective Diagnostic Tests

Harry Dill
Deep Creek Technologies, Inc.

Keywords: SPICE, fault simulation, diagnostics, diagnostic inferencing, testing, pass/fail limits, fault detection, fault isolation, test program sets.

Abstract: This chapter addresses the role of Pass/Fail limits in the detection and isolation of faults in electronic circuits. Emphasis is placed on the use of Pass/Fail limits to discriminate among suspected failures in a known faulty circuit and improving test outcome confidence. Analog circuit simulations serve to illustrate failed circuit behavior characteristics and the effect of assigning multiple Pass/Fail limits to a single test setup for improved diagnostic resolution.

1.   INTRODUCTION

One of the most difficult tasks of test synthesis is definition of the criterion that will be used to assign a Pass or Fail outcome to a test. This criterion, more commonly known as Pass/Fail limits, is the principal factor governing key test attributes such as probability of false alarm and probability of detection. Pass/Fail limits also have a profound influence on the diagnostic inferences provided by a test. Ultimately, the Pass/Fail limits influence the quality of diagnostic test sequences. This chapter examines new, structured techniques for defining Pass/Fail limits with the objective of creating diagnostic fault trees that isolate failures to smaller ambiguity groups with higher confidence levels.

The experimental data used to derive and support the concepts presented in this chapter were formulated from behavioral analysis of a Motor Current Transducer Input Circuit found in a rail transportation car.

The circuit schematic is shown in Figure 1. All behavior analyses were prepared using a SPICE based simulator.

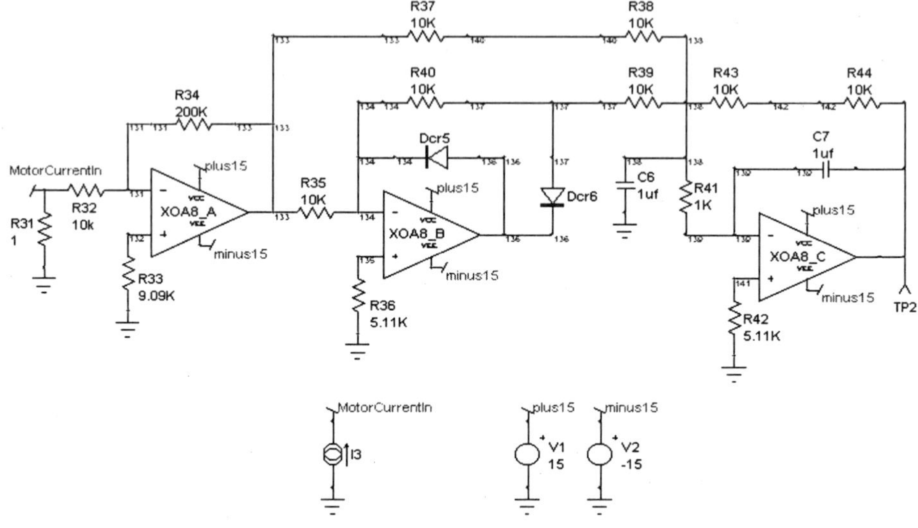

Figure 1 - Motor Current Transducer Input Circuit Test Objectives

2.   TEST DEFINITION

There are many different reasons for performing tests. The proof of a hypothesis, evaluation of the performance of a new design under environment extremes, determination of reliability characteristics and verification of fault free performance are but a few. This chapter concerns itself with the use of tests to detect and isolate failures in analog electronic circuits of proven design (Boros, 1985).

All tests used for detection and isolation of faults in an analog circuit are defined by four distinct characteristics—stimuli, observed parameter, point of observation and outcome criteria.

- The stimuli define the environment that the Unit Under Test (UUT) will be exposed to prior to or during the observation of a parameter. They can be as simple as a light source to illuminate components during a visual check for discoloration or as complex as thousands of frequency/time division multiplexed RF signals applied to multiple UUT inputs.
- The observed parameter is what is to be measured during the test. Although electronic circuit measurements usually involve voltages, currents, impedances, time and occasionally temperature; measurements may be less quantitative and include odor and color.

- The location of the observation is where on the UUT the observation is to be made. If the observed parameter is component color, the observation point is the location of the component. If voltage is the observed parameter, the observation point is the circuit node where the voltage is of interest.

- The outcome criterion is the standard against which the observation will be compared to determine what information can be inferred from the test outcome. The criterion is usually quantitative in nature and is defined as a "spread" of values. An example of a test outcome criterion might be "if a voltage at node X is between 4.5 Volts and 5.5 Volts, the power supply circuit is fault free, otherwise it is faulty." The outcome criteria does not have to be quantitative. The test might be "does the circuit board smell burnt?" In this case, the accuracy of the instrument will vary widely with the tester. Usually only two outcomes are defined for a test—Pass and Fail. However, since Pass and Fail are each associated with some range of an observed parameter and the range of the observed parameter can be subdivided into any number of ranges, the methodologies discussed herein can be easily extended to multiple outcome tests (e.g., Fail Hi, Pass, Fail Lo).

Change any one of these four characteristics and the inferences that can be derived from the test outcome may also change.

The testing objective determines the Pass/Fail limits definition approach. There are many possible testing objectives but the two prominent objectives of industry are *performance verification* and *fault isolation*.

2.1   Performance Verification

Performance verification tests are used to confirm that an assembly is fault free. The most common point of application is on the assembly line as fault isolation and repair is usually not of interest. The objective is to ensure that all assemblies are operational and ready for installation at the next higher level of assembly.

Definition of Pass/Fail limits for performance tests is usually accomplished using one of two methods:

1. The limits are derived from the parametric spread limitations of the inputs at the next higher level of assembly, or
2. The tolerances of each of the components of the Unit Under Test (UUT) are used in a Monte Carlo statistical computation to determine limits based on the probability of passing an assembly with a failure in it and failing an assembly with no failure.

Method 1 is the easiest approach *if specifications of the input requirements for the next level of assembly are available.* Often they are not available. For example, a DC Power Supply vendor rarely knows the exact requirements of all the customers. The customer uses a supply if it meets *or exceeds* requirements.

Method 2 can present unnecessary testing challenges as parametric variations of the UUT outputs that result from the use of commonly available components are often much narrower than the design tolerances required at the next higher level of assembly. Consider the example circuit in Figure 1. The resistors each have a tolerance of 5%. The capacitors each have a tolerance of 10%. By performing a Monte Carlo analysis on the example circuit, it is possible to compute the expected value and standard deviation of the voltage at each node (Staudhammer, 1975). Table 1 shows the mean and 3 sigma voltage values for the example circuit given a Motor Current stimuli of a sine wave voltage having frequency of 71.5 Hz and peak amplitude of 0.3 Volts.

Tests have been defined for each node in the circuit. The measured parameter for these tests is mean voltage. Pass/Fail limits have been defined at ± 3 sigma about the mean. This yields a 1.3×10^{-3} probability of failing any of the tests defined for a known good Motor Transducer Input assembly. Note that the spread between the upper and lower Pass/Fail limit (2 x 3 sigma) is less than 50 mv for 11 out of the 14 tests. This limit spread is below the noise floor of many large automatic test systems (~50mV) and makes 10 of the 14 tests unusable. However, this spread may be much smaller than necessary if consideration is given to the failure mechanisms of the circuit (more on this in the next section). Consider that given no information about failed behavior of the circuit, any measurement outside of the Monte Carlo derived limits indicates that the circuit is faulty. Use of tests having such narrow limits places severe constraints on test instrumentation because instrumentation error and test system noise could easily invalidate the results and identify a fault free UUT as failed.

2.2   Fault Isolation

Fault isolation tests are used to identify a faulty component within the UUT. The most common point of application is the maintenance and repair facility. The objective is to locate the faulty component, make repairs, and return the UUT to service.

Table 1 - Node voltage variations due to component parameter in-tolerance variations.

Measurement Node	Mean Voltage	3 Sigma Voltage
131	11.08 uV	82.17 nV
132	-409.9 nV	18.97 nV
133	-38.01 mV	12.55 mV
134	19.45 uV	3.180 uV
135	-230.8 nV	10.53 nV
136	-1.913 V	0.1926 V
137	-1.895 V	0.1890 V
138	115.1 uV	264.6 uV
139	-6.295 uV	2.246 uV
140	-18.99 mV	6.169 mV
141	-229.9 nV	9.336 nV
142	1.926 V	0.2546 V
MotorCurrentIn	1.902 mV	595.8 uV
TP2	3.853 V	0.5098 V

Fault isolation tests offer a broader latitude for limit setting as their "Pass" outcome does not necessarily require that there is no failure in any UUT component involved in the generation of the observed parameter. As a simple illustration, Figure 2 shows the spread of measurements at node 138 of the example circuit when there are no failures. The X-axis of Figure 2 has the units of Volts and has been expanded around the 4.75 sigma "no fault" range to illustrate just how small this range is. Plotted together with this "no failure" measurement spread are the nominal values of node 138 voltage for the predefined failure modes. There are four failure modes contained within the "no fault" range that are not detected by this test regardless of the Pass/Fail limits. The simulation for this example uses failure mode default values of 100 megohms for open resistors, open diodes and open capacitors, and 0.1 ohms for shorted capacitors.

Figure 2 also shows a single Pass/Fail limit that has the objective of isolating failure mode R44 Open from some (but not all) of the other suspect failure modes. If the test fails (i.e., is below the limit), failure mode R44 Open may have been the culprit; if the test passes, there is high confidence that failure mode R44 Open is not a problem. Note that the pass region for the fault isolation test shown in Figure 2 encompasses all of the expected no fault range of values (2 x 4.75 sigma) as well as a large portion of the fail range of voltages for the circuit shown in Figure 1. If the outcome is defined as Fail for voltage observations below the criterion and Pass for observations above the criterion we can use the Fail outcome to isolate R44 Open, R43

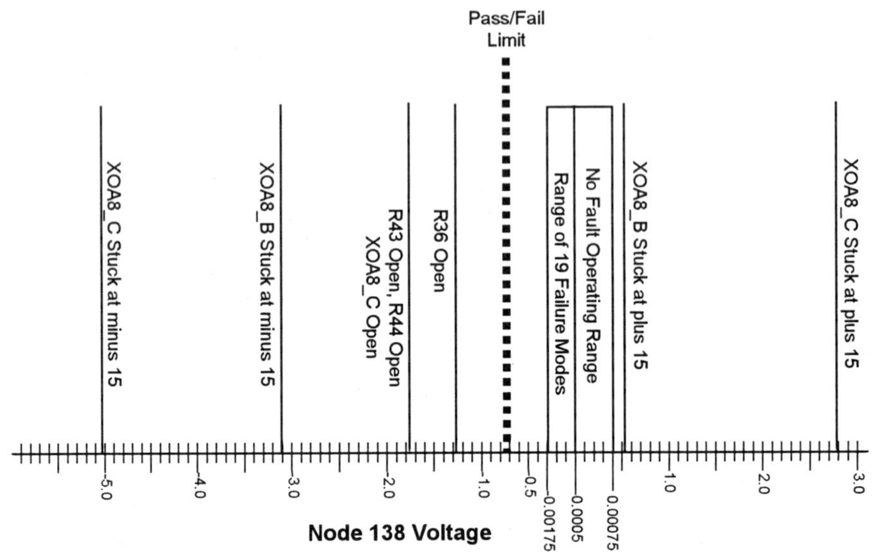

Figure 2 - Node 138 Voltages For Failure Modes and No Failure

Open, R36 Open, XOA8_C stuck at minus 15, XOA8_C Open and XOA8_B stuck at minus 15 from the other 25 failure modes in the circuit.

Placing a Pass/Fail limit at the -0.75 Volt level for the example circuit offers several advantages over placing Pass/Fail limits at the 4.75 sigma voltage levels about the mean.

- The test outcome is more reliable because the spread between the upper and lower Pass/Fail limits is much greater than the noise floor of the test system. Actually, for this example there is no upper limit; as long as the noise floor of the test system plus the lowest expected measurement in the "Pass" range is greater than the Pass/Fail limit, the probability of false failures due to system noise will be low.

- The test inferences are more reliable. Consider the case of ±3 sigma Pass/Fail limits for the example circuit. Figure 2 shows 19 failure modes with corresponding node 138 voltages within 161 millivolts of the lower Pass/Fail limit. Five of these failure modes have associated node 138 voltages less than 1 millivolt from the lower limit. Because each circuit card in a population has its own unique set of tolerances, UUTs having node 138 voltages at the high end of the distribution for the case of no failures may shift the X-axis of Figure 2 sufficiently far to the left relative to the fixed Pass/Fail limits to make several of the 19 failure modes undetectable. The -0.75 Volt Pass/Fail limit shown in Figure 2 has a much higher immunity to this inference uncertainty. There is a 500 millivolt "buffer zone" on both the high and low sides of the limit. If the test fails, there is high confidence that the culprit is R44 Open, R43 Open, R36 Open, XOA8_C stuck at minus 15, XOA8_C Open or XOA8_B stuck at minus 15. If the test passes there is high confidence that the culprit is one of the other 25 failure modes or the circuit contains no failures.

- Tests having "tailored" failure mode inferences can be synthesized by moving the Pass/Fail limits to the appropriate position on the measurement range. For example, consider a Pass/Fail limit placed at the +1.5 Volt level in Figure 2 and assign an outcome of "Fail" to the test if the observed parameter is greater than this limit. By doing nothing more than moving the Pass/Fail limit from -0.75 Volts to +1.5 Volts, the failure mode inference has been changed from R44 Open, R43 Open, R36 Open, XOA8_C stuck at minus 15, XOA8_C Open and XOA8_B stuck at minus 15 to XOA8_C stuck at plus 15.

3.   UUT BEHAVIOR CHARACTERIZATION

In order to take advantage of failure information during the development of effective performance and fault isolation tests, the behavior of the circuit must be characterized in terms of what parameters will be observed at each observation point for each failure mode. Pass/Fail limits can then be assigned to achieve the desired fault detection (e.g., inference) characteristics. Until recently, this task was computationally intractable for all but the simplest of circuits. With modern computer clock speeds, simulation has become a viable solution.

3.1   Failure Free UUT Behavior

There are three views of failure free UUT behavior—each correct when taken in context.

1. *UUT meets all manufacturer's performance specifications.* For example, a DC Power Supply may have requirements of 5±0.001 Vdc for a load of 1 amp and an input of 120±15 Vac at 60 Hz.
2. *UUT meets all performance requirements for the next higher level of assembly.* A digital control circuit with discrete outputs may require power of 5±0.5 Vdc for a load of 250 ma and an input voltage of 120±15 Vac at 60 Hz. Leaky filter capacitors may render the DC Power Supply cited in view 1 incapable of meeting manufacturers specifications. That same "failed" Power Supply may be fully functional in the context of the digital control application.
3. *A set of tests designed to detect all UUT failures in a predefined UUT failure universe show that no failures in that universe exist.* This view is often used by test engineers when designing fault

isolation tests that have measurements at observation points for which no manufacturer's specifications exist and the measured parameters are not tightly coupled to the requirements at the next higher level of assembly.

The approach presented here combines these three views by testing UUT outputs to the next higher level of assembly requirements if they are known and the manufacturer's specifications if the next higher level of assembly requirements are not known. The view 3 approach is applied when the available instrumentation cannot accurately perform measurements within the UUT manufacturer's tolerances and the next higher level of assembly requirements are unknown.

Figure 3 shows the TP2 output of the example circuit along with the nominal expected voltage for each failure mode. Also shown in the figure are 4.75 sigma Pass/Fail limits. These limits represent a 1.62 Volt spread. To illustrate the use of failure mode behavior information for increased test outcome confidence, consider the hypothetical case where the voltage at TP2 is to be measured with an instrument having a three sigma accuracy of 1.5 volts and a Gaussian error distribution. With Pass/Fail limits set at 1.62 Volts, the instrumentation error would cause the test outcome to occasionally Fail when in fact the TP2 voltage is within the 1.62 Volt Pass range. This incorrect fail outcome is commonly known as a false alarm. (The frequency of false alarm is a function of the instrument accuracy and other factors.) Now consider the case when the Pass/Fail limit spread is increased to 4 Volts (±2 Volts about the mean). An examination of Figure 3 shows that there is no effect on the fault detection characteristics of the test. However, the number of false alarms for this test should decrease as the Pass/Fail limits are now significantly further out on the tails of the instrument error distribution (Bolten, 1992).

3.2   Failed Behavior

Computation of failed circuit behavior requires knowledge of component failure mode parameters. For purposes of this chapter, component failure mode definitions have been extracted from Navy standards and failed component parameters have been assigned to every component in the example circuit (e.g., open resistor is 100 megohms, shorted transistor collector-emitter is 0.1 ohm, etc.). This set of failures is known as the failure universe and represents the set of conclusions (with the addition of No Fault) from which we must draw during fault isolation.

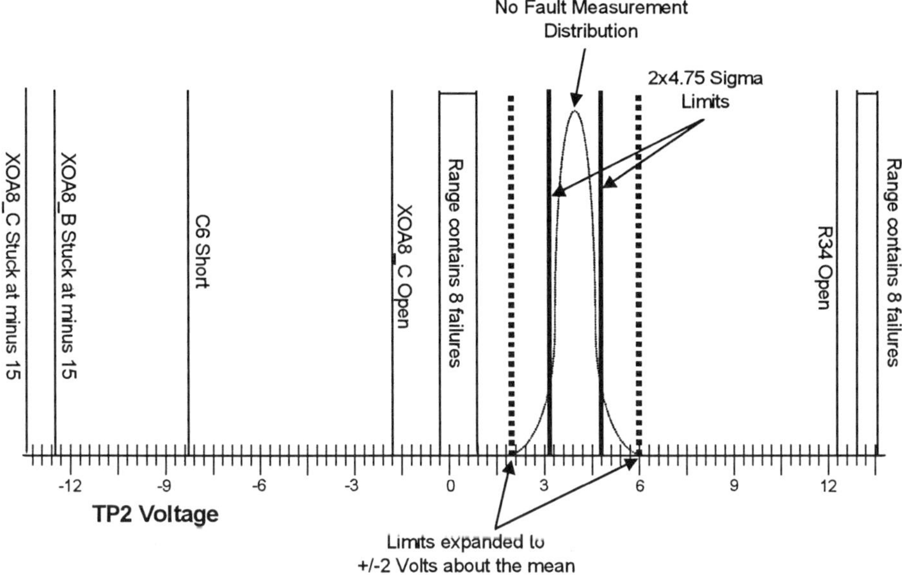

Figure 3 - Failure information shows limits can be expanded without affecting test results.

Figure 2 shows the failed voltage measurement to be expected at node 138 of the example circuit for each failure mode parameter defined in the fault universe. Note that we are defining a specific value for the component failure mode parameter in this analysis. In the real world, a spread of failed values is a more realistic assumption, however, a Monte Carlo analysis around every failure mode parameter is computationally prohibitive. Therefore, when capturing failed behavior for use in setting test limits, it is incumbent upon the test engineer to examine each failure mode definition and determine the most suitable parameter(s) for failure mode characterization. Some failure modes drive circuit parameters into stable states over wide ranges of component failure parameters. Other failure modes drive circuitry into non-linear operational regions that make the circuit ultra sensitive to variations in the parameter used for the failed component. Fortunately, modern circuit simulators automate analysis of component parameter variations and facilitate selection of appropriate component failure parameters.

An interesting sidenote: Experience with the SPICE simulator has shown that for some failure modes, a circuit will exhibit instability for the stimuli of Test A and the failed circuit behavior is unpredictable; yet, for the same failure modes, circuit stability is excellent for the stimuli provided by Test B and the failed circuit behavior is highly predictable. This argues for *ordering tests to eliminate failure modes from consideration that cause circuit instabilities in subsequent tests.* This is a subject for future investigation.

4.   SETTING PASS/FAIL LIMITS

The previous sections show how individual failure modes induce specific circuit behavior that can be characterized for a given test stimuli. This information can be captured in tests and used to decrease the granularity of fault isolation and increase the ability to reliably discriminate among failure modes. This section of the chapter describes the creation of multiple tests that use exactly the same stimuli, measured parameter and measurement location. These tests differ *only* in their respective Pass/Fail limits definition.

4.1   Binary Tests

Binary tests are defined as tests having an upper Pass/Fail limit and a lower Pass/Fail limit. If the measured parameter is greater than the lower limit and less than the upper limit, the test outcome is defined as Pass, otherwise, the outcome is Fail. This is the conventionally implemented test type for performance testing. It assumes that if the outcome is Pass, then no failure mode was detected.

Now let's take an unconventional view of the binary test. Figure 4 shows a graph of the TP2 voltage versus failure modes for the example circuit. Note that the Pass/Fail limits have been intentionally set such that the Pass region (e.g., the range of parameters between the lower and upper limit) encompasses parameters expected for nine failure modes but excludes the parameter range for No Fault. Also note that the Fail regions (e.g., measurements greater than the upper limit and less than the lower limit) include not only expected measurements corresponding to failure modes but also the range of parameters defined for No Fault. Thus, by setting limits, we have defined the inferences (e.g., fault detection characteristics) that can be made from the Pass outcome (nine failure modes) and the Fail outcome (22 failure modes plus No Fault) of this test. By moving the limits around in this manner, the test engineer can tailor the detection characteristics of the test to the fault isolation requirements. Applying this technique, it is possible to design a set of tests using a single set of stimuli and a single measurement at a given point, that have enhanced fault isolation characteristics over binary tests that center Pass/Fail limits around the No Fault parameter range.

4.2   Tertiary Tests

Tertiary tests are an extension of binary tests in that they consider only one Pass/Fail limit at a time. The measured parameter can be greater than the upper limit (Test Outcome is Fail Hi), less than the lower limit (Test Outcome is Fail Lo), or within the Pass region bounded by the lower and upper limits. The inferences that can be made from the tertiary test are:

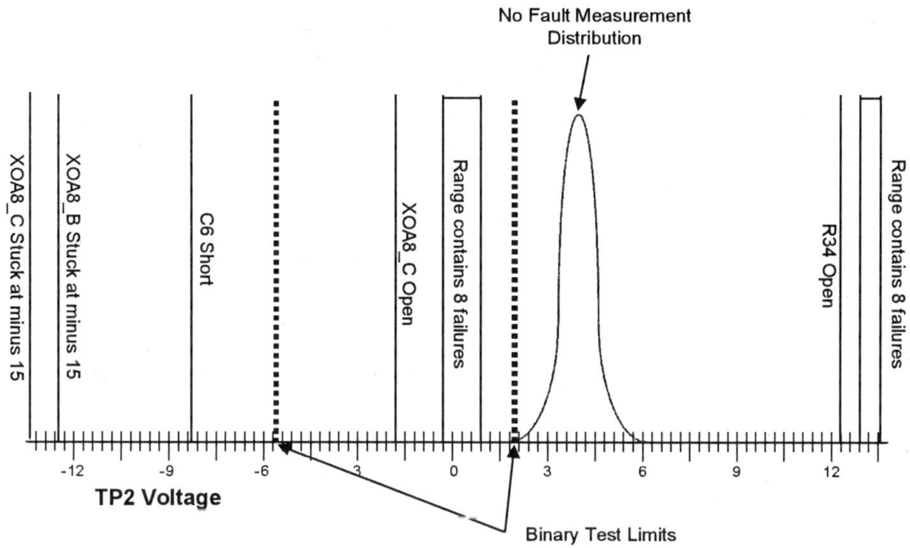

Figure 4 - Binary test Pass/Fail limits set for specific fault detection characteristics.

1. If the test outcome is Fail Hi then the failure must be in the set of failures associated with the parameters above the upper limit
2. If the test outcome is Fail Lo then the failure must be in the set of failures associated with the parameters below the lower limit.
3. If the test outcome is within the Pass region, the failures associated with parameters above the upper limit and below the lower limit are eliminated from consideration.

The Pass/Fail limits defined for the binary test shown in Figure 4 can also be used for a tertiary test. In this example, if the test has an outcome of Fail Lo then three failure modes are inferred - C6 Short, XOA8_B Stuck at -15 Volts and XOA8_C Stuck at -15 Volts. If the test outcome is Pass, XOA8_C Open 8 other failure modes and No Fault are inferred. If the test outcome is Fail Hi, 20 failure modes are inferred.

5.   SUMMARY

The parameters that characterize a failed circuit component govern the behavior of the circuit containing the failure. Knowledge of circuit response to component failure modes can be used during the synthesis of diagnostic tests to control the inferences that can be made for each test outcome. The specific mechanism is the selection of Pass/Fail limit criteria.

Knowledge of circuit response to component failure modes is useful for increasing the reliability of test outcomes. By setting Pass/Fail limits in regions away from expected failed parameter values and no failure parameter values, the probability of false alarm is reduced.

6.   REFERENCES

Bolten, W., Electrical and Electronic Measurement and Testing, Longman Scientific and Technical, 1992, pp. 1-21 and 235-246.
Boros, A., Electrical Measurements In Engineering, ELSEVIER, 1985, pp. 29-40.
Staudhammer, John. Circuit Analysis By Digital Computer, Prentice-Hall, Inc., 1975, pp. 292-307.

Chapter 3

Fault Hypothesis Computations Using Fuzzy Logic

Timothy M. Bearse, Michael L. Lynch
Naval Undersea Warfare Center

Keywords: Fuzzy logic, model-based diagnostics, diagnostic inference model, diagnostic reasoner, test inference uncertainty, test outcome uncertainty, fault hypothesis, diagnostic reasoning conflict, AI-ESTATE.

Abstract: The authors apply fuzzy logic techniques to fulfill the requirement to compute a fault hypothesis in a model-based diagnostics reasoner. The primary focus of the research is to illustrate the use of fuzzy sets to derive the current fault hypothesis given test outcomes. In the approach, measurement parameters are mapped to test outcomes using membership functions, and a fuzzy logic processor is used to combine the evidence associated with multiple outcomes from a test as well as information from multiple tests. The chapter deals with two types of uncertainty: test outcome uncertainty and test inference uncertainty. The issues of conflict detection and mitigation, and rehabilitation of conclusions are addressed.

1.   INTRODUCTION

When performing test and diagnostics using a model-based diagnostics approach, the process has uncertainty associated with it that cannot be avoided. During the model development process, inferences are determined using engineering judgment and/or simulation. Due to the computational cost of fault simulation, the inferences are developed using as few simulations as possible. As such, these inferences aren't always representative of the entire population of circuits, especially with faults inserted. In addition, the inferences are based on test outcomes, not the measurement itself. A measurement is mapped to a specific test outcome and then the inferences associated with the resulting test outcome are assigned truth values. There is uncertainty associated with the measurement itself since instruments have tolerances associated with them and measurement systems are noisy.

An analysis of Bayesian inference, Dempster-Shafer inference, fuzzy logic and the MYCIN/EMYCIN calculus was presented as potential reasoning under uncertainty mechanisms applied to medical diagnosis (Harrison and Henkind, 1988). Uncertainty calculi such as fuzzy logic and Dempster-Shafer theory have been proposed for use as reasoning under uncertainty mechanisms in model-based diagnostics (Sheppard and Simpson, 1991a). In that work, the application of fuzzy logic was mostly in the user interface and was not the primary inferencing mechanism. This work presents an application of fuzzy logic as the primary reasoning under uncertainty mechanism for a model-based diagnostics reasoner.

1.1   Model-based Diagnostics Background

The use of dependency models and Diagnostic Inference Models (DIMs) to determine the testability of systems as well as to diagnose systems is well-documented in the literature (Sheppard and Simpson, 1991b; Sheppard, 1994; Simpson and Sheppard, 1994; Dill, 1994). These types of models form the basis for IEEE Std 1232.1:AI-ESTATE (1997). The original forms of these models were based solely on concepts founded in propositional logic. Typically, classical set theoretic operations are performed on these models (Sheppard and Simpson, 1991b; NUWCDIVNPT, 1994) in order to determine the fault candidates which exist given specific test outcomes.

The IEEE AI-ESTATE committee extended the definition of the DIM to include confidence information: test outcome confidence and test inference confidence. Due to the inevitable limitations of processing models limited to two truth values for test outcomes and failure modes, this decision is prudent. This confidence data can be made available for use by various AI-ESTATE compliant reasoners during test and diagnosis of a UUT. This chapter addresses the use of these confidence parameters.

1.2   Fuzzy Logic Background

In classical set theory, an individual object, x, either is or is not a member of set A. The crisp set is defined in order to partition the individuals in some universe of discourse, X, into one of two groups: members (those that certainly do belong) and non-members (those that certainly do not belong) (Klir and Yuan, 1995). There are primarily three methods to define sets: the list method, the rule method and the characteristic function method. In classical set theory, a characteristic function maps the elements of X to elements of the set $\{0,1\}$. There are no provisions for partial membership in a set, hence there is no inherent mechanism for dealing with uncertainty.

Fuzzy set theory (Klir and Yuan, 1995) allows for the assignment of a value which represents the grade of membership of each individual in the given universe of discourse in a specific fuzzy set. The value determines the degree to which the individual is a member of the fuzzy set. The mechanism maps the elements of X to elements in the interval [0,1]. The mapping is performed by a membership function. The fuzzy set defined by the membership function must have some meaning in the problem domain and the degree of membership must be interpreted relative to the fuzzy set and the domain. For example, fault candidate can be defined as a fuzzy set. Each failure mode in the fault universe can be mapped to the fault candidate fuzzy set. In crisp set theory, each failure mode could take on one of two values: 0 or 1. In fuzzy set theory, the failure mode can be assigned a value in the interval [0,1]. This value would then represent the degree of membership that the failure mode is a fault candidate or the degree of truth in the statement that the failure mode is a fault candidate.

1.3   Diagnostic Fuzzy Variables and Subsets

In traditional processing of DIMs, a specific test outcome is considered to have occurred or not occurred. There is no inherent mechanism to process information that isn't completely certain. Using the fuzzy logic paradigm, the outcome of the test may be defined as a fuzzy variable and represents the state of the test. States of the fuzzy variable are fuzzy sets representing the specific outcome confidences of the test. For example, states of the fuzzy variable Test Outcome could be defined for a three outcome test as: Meas. Low, Meas. Nominal and Meas. High. Each state or specific outcome is a fuzzy set and has a representative membership function depicted in Figure 1. In a crisp world, the Test Outcome variable would be assigned either Meas. Low, Meas. Nominal or Meas. High. In the fuzzy world, Test Outcome variable can have membership in all three states. Each specific test outcome is considered to be a fuzzy subset defined on the interval [0,1]. In this manner, the partial truth of the fact that a specific test outcome occurred can be represented. It is the degree of truth in the statement that the test outcome occurred and is defined below as test outcome confidence.

Likewise, in traditional DIMs, a failure mode either exists or does not exist in a fault hypothesis. There is no way to discuss the partial truth of the statement that a failure mode exists. In the fuzzy logic based approach, the concept of fault candidate is defined to be a fuzzy subset defined on the interval [0,1]. It is the degree of truth in the statement that a specific failure mode is a fault candidate.

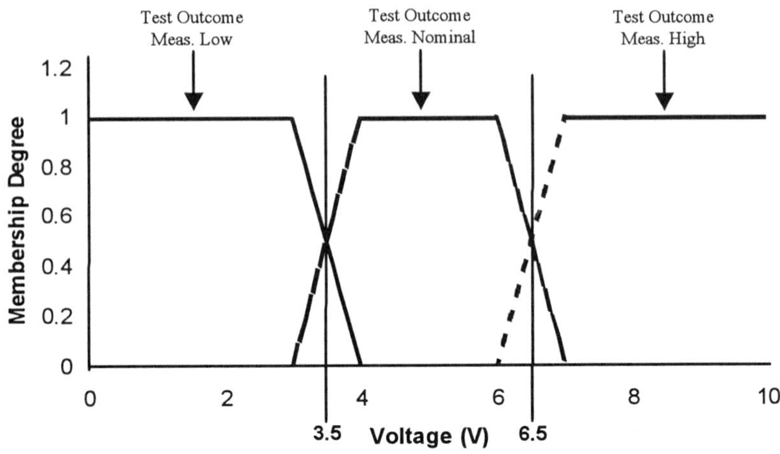

Figure 1. Membership Functions For Outcomes Of Test, T1

2. TEST OUTCOME CONFIDENCES

Test outcome uncertainty is represented by the test outcome confidence parameters which are defined using the test outcome fuzzy subsets. The membership value in a specific test outcome represents the confidence in that test outcome. There are N outcome confidence parameters, one for each of the N outcomes for a test. Each outcome confidence parameter can be either provided by the user (a priori or on-line) or mapped from the measurement parameter to specific test outcomes/outcome confidences for that test using membership functions after the execution of the test. If the test outcomes are mapped from the specific measurement parameter at run-time, an N-outcome test would need N membership functions. The test can be defined by the N membership functions:

$$TC_i : TestResult \rightarrow [0,1]$$

where $i = 1,..., N$. A specific measurement parameter could then result in partial membership in different outcome subsets for the same test and a confidence is generated for each test outcome.

In a crisp world, consider test T1 which has three possible outcomes:

Meas. Low: $0V \leq voltage < 3.5V$

Meas. Nominal: $3.5V \leq voltage < 6.5V$

Meas. High: $6.5V \leq voltage < 10V$

In this crisp world, if the measurement was 3.2 volts, then the test outcome would simply be Meas. Low. Meas. Low would be assigned to the crisp variable Test Outcome.

In the world of fuzzy sets, membership functions associated with the test outcomes can be used. Figure 1 depicts three membership functions for test T1 which has three possible outcomes: Meas. Low, Meas. Nominal and Meas. High. For a measurement of 3.2 volts, the following membership values would be assigned:

> Meas. Low: 0.8
> Meas. Nominal: 0.2
> Meas. High: 0.0

This assignment means that there is 0.8 degree of truth to the statement that the test outcome was Meas. Low, 0.2 degree of truth to the statement that the test outcome was Meas. Nominal and 0.0 degree of truth to the statement that the test outcome was Meas. High given the measurement of 3.2 volts. As seen, a specific measurement can result in membership in more that one specific outcome. Fuzzy sets allow for dealing with the uncertainty associated with the test outcome in an intuitive manner. As the measurement parameter approaches the boundary between two different test outcomes, the crisp approach forces a hard decision to be made. This approach does not allow for the real world situation that there is uncertainty associated with the measurement. The fuzzy set theoretic approach allows for the natural assignment of membership in more than one outcome for a specific measurement parameter.

3.   TEST OUTCOME AND FAULT INFERENCE CONFIDENCES

In traditional DIMs, the relationship between a test outcome and the faults that may be suspect if that test outcome occurs is presented as an implication list. One implication list is created for each test outcome in the model. The model consists of a set of implication lists, one implication list for each test outcome. The processing of these lists usually assumes that test outcomes are certain and that the inferences associated with each outcome are certain.

The test-fault inference confidence is the parameter which captures the notion that the inference between a test outcome and a specific fault is not certain. A test outcome-fault inference confidence is associated with the inference, $T1 \rightarrow F1$, and represents the degree of truth associated with a fault given the test outcome occurred, where T1 is the fact that test outcome T1 occurred and F1 represents the fact that failure mode F1 is a fault candidate. This inference confidence can be represented as:

$$T1 \xrightarrow{\text{fc}} F1$$

where fc represents the inference confidence parameter.

4. QUANTITATIVE MODEL FOR DIAGNOSTICS

Formally, the model that we will be discussing in this chapter consists of:

- set of *n* tests $T = \{t_1, \ldots, t_n\}$
- set of *N* Test Outcomes $TO = \{TO_1, \ldots, TO_N\}$, where each Test Outcome, TO_x, is a fuzzy variable and has N fuzzy subsets associated with it.
- matrix of *n* by *N* test outcome confidences:
  $$TC = \{TC_{11}, \ldots, TC_{1N},$$
  $$TC_{n1}, \ldots, TC_{nN}\}$$
  where each TC_{ij} can be determined using an appropriate membership function $TC_{ij} : \text{MeasurementParameter}_i \rightarrow [0,1]$ or assigned by the user.
- a set of *m* faults $F = \{f_0, f_1, \ldots, f_{m-1}\}$
- a *m by n by N* matrix of test outcome-fault inference confidences, $FC = \{fc_{ijk}\}$, where i is the fault index, j is the test index and k is the outcome number for the jth test.
- a set of *m* candidate membership values $FH = \{c_0, c_1, \ldots, c_{m-1}\}$. This set represents the fault hypothesis. Each c_i represents the degree of membership that the $fault_i$ is a candidate for failure. FH_c represents the current fault hypothesis and FH_{Tx} represents the candidate membership values after only considering the test outcomes associated with test T_x.

Without loss of generality, it is assumed that each test has N outcomes associated with it.

5. QUANTITATIVE MODEL PROCESSING

The objective is to determine the fault hypothesis by computing the degree of truth that each fault is a candidate given the outcomes of tests and the quantitative model for the system. The outcome of the tests and the test outcome-fault inferences associated with each outcome can be uncertain. The process to compute the fault hypothesis consists of the following steps:

- Determine the uncertainty associated with the executed test
- Determine candidate membership degree for all fault candidates for current test
- Combine current membership degree with membership degree for current test
- Normalize each membership value in the candidate fuzzy set
- If desired, threshold the membership values in the candidate fuzzy set

5.1 Determine Uncertainty Associated with Test

In order to determine the uncertainty associated with the test, test outcome confidence parameters must be provided. These parameters may be assigned by the user at run-time or during model development, or generated by mapping the measurement parameter to test outcome confidences via membership functions at run-time.

User Assigned Confidences. The model developer or the user of the reasoner supplies the base confidence, TC_B, for a specific test outcome. The base confidence can be supplied during model development or at run-time. The remaining outcome confidences for the test can be provided by the user or computed using $(1-TC_B)/(N-1)$, where N is the number of outcomes for the test. The resulting set of test outcome confidences represents the uncertainty of the test.

Membership Functions. For an N-outcome test, N membership functions can be created which map the measurement parameter to specific test outcome confidences. Figure 1 provides an example of a set of membership functions for a three outcome test. Once we have test outcome confidences $\{TC_{j1}, \ldots, TC_{jN}\}$ for the current test, t_j, we can determine the candidate membership values. The fuzzy variable TO_j is assigned values $\{TC_{j1}, \ldots, TC_{jN}\}$.

5.2 Determine Candidate Membership Given Test

Let the fault hypothesis represent the set of membership values for all faults in the candidate fuzzy subset, $FH = \{c_o, \ldots, c_{m-1}\}$. In order to compute the membership degree associated with each failure mode in the fault hypothesis, both the test outcome confidence and the test-fault inference confidences must be considered. There are a number of ways that the confidences can be combined.

The following compositional rule of inference is used to determine the degree of truth that the failure mode is in the candidate set for a given test, t_j:

$$C(f_i) = \max_{to \in TO_t} \{\min[TO_t(to), R_i(to, f_i)]\}$$

for each $f_i \in F$

where: R_i is a binary fuzzy relation between the test outcomes, to, of the fuzzy set, TO_t and the faults, f_i. This relation is obtained from the test outcome inference confidences matrix for specific fault index, i, and test index, j. The fuzzy set C represents the degree of truth in the statement that f_i is a fault candidate. Each fault has a membership grade, possibly zero, in this candidate fuzzy set.

When the confidence associated with a specific test outcome is less than one, the inferences associated with the other test outcomes of the same test are taken into consideration. It is also possible to use other operators (i.e., *product*) instead of the *min* operator in the equation above. This operator was selected based on the fact that it is so common in the literature.

<u>An outcome-based scenario:</u>

1. Get crisp test outcome, k, for executed N-outcome test, t_j.
2. Fetch the base confidence value (user or model supplied) for this test outcome (TC_B)
3. Compute TC_{jx} for remaining outcomes using $(1-TC_B)/(N-1)$
4. For each fault i, compute candidate membership set, c_{ij}, by applying the *min* operator on $[TC_{jk}, fc_{ijk}]$ for each test outcome k of test t_j.
5. For each fault i in universe, determine the candidate membership value using: $c_i = \max \{c_{ij}\}$
6. $FH_{tj} = \{c_o, \ldots, c_{m-1}\}$

<u>A measurement-based scenario:</u>

1. Get measurement for test, t_j.
2. Apply measurement to each of N test outcome membership functions associated with the test. This will result in a vector of test confidences $\{TC_{j1}, \ldots, TC_{jN}\}$.
3. For each fault i, compute candidate membership set, c_{ij}, by applying the *min* operator on $[TC_{jk}, fc_{ijk}]$ for each test outcome k of test t_j.
4. For each fault i in universe, determine the candidate membership value using: $c_i = \max \{c_{ij}\}$
5. $FH_{tj} = \{c_o, \ldots, c_{m-1}\}$

If the confidence of any test outcome is one (certain test outcome), then the candidate set, the fault hypothesis after considering t_j, will consist of the test outcome-fault inference confidences, c_{ij}, associated with that test outcome. In addition, if the matrix of test outcome-fault inference confidences consists entirely of ones and zeroes, then the inferences are certain and the result is the same fault hypothesis that would be obtained using traditional set theory.

Example. Consider the binary outcome case with three (3) tests, four (4) conclusions and model data in Tables 1 and 2. The problem is to determine the set of candidate membership values associated with T0 when the test has an outcome of pass.

Table 1. Inference Confidence Data

	T0-fail	T0-pass	T1-fail	T1-pass	T2-fail	T2-pass
F0	0.01	0.99	0.01	0.99	0.02	0.98
F1	0.98	0.02	0.02	0.98	0.97	0.03
F2	0.98	0.02	0.97	0.03	0.03	0.97
F3	0.03	0.97	0.98	0.02	0.98	0.02

Table 2. Test Outcome Base Confidence Data

	Test Outcome Confidence
T0-pass	0.8
T0-fail	0.9
T1-pass	0.95
T1-fail	0.95
T2-pass	1
T2-fail	1

Assuming for the moment that T0–pass has a confidence of one, the fault hypothesis after the test was executed and, in fact, passed would be {(F0, 0.99), (F1, 0.02), (F2, 0.02),(F3, 0.97)}. There would be no need to consider the impact of the T0–fail outcome since the T0–pass outcome is considered to be certain. This is the usual case when traditionally processing DIMs.

However, as we can see from Table 2, T0–pass has a test outcome confidence of 0.8. The potential impact of test outcome uncertainty is a reduction in the degree of truth associated with that fault. Since the pass test outcome confidence is only 0.8, one should additionally consider the impact of the T0–fail outcome on the fault hypothesis since we are not completely certain that test T0 passed.

If we have a 0.8 confidence that test outcome was pass, it is reasonable to assume that we have a 0.2 confidence that the test outcome was fail. We will follow the procedure for the outcome-based scenario to determine the candidate membership grades.

1. Get crisp test outcome for executed binary outcome test, T0:

    $T0 = pass$

2. Look up base confidence value for this test outcome, TC_B:

    $TC_B = TC_{11} = 0.8$

3. Compute TC_{xj} for remaining outcomes using $(1-TC_B)/(N-1)$:

    $TC_{12} = 0.2$

4. For each fault i, compute candidate membership set, c_{ij}, by applying the *min* operator on $[TC_{jk}, fc_{ijk}]$ for each test outcome k of test t_j:

    $c_{01} = \{min[TC_{11}, fc_{011}], min[TC_{12}, fc_{012}]\}$
    $c_{01} = \{min[\ 0.8, 0.99], min[0.2, 0.01]\}$
    $c_{01} = \{\ 0.8, 0.01\}$
    $c_{11} = \{min[TC_{11}, fc_{111}], min[TC_{12}, fc_{112}]\}$
    $c_{11} = \{min[\ 0.8, 0.02], min[\ 0.2, 0.98]\}$
    $c_{11} = \{0.02, 0.2\}$
    $c_{21} = \{min[\ TC_{11}, fc_{211}], min[\ TC_{12}, fc_{212}]\}$
    $c_{21} = \{min[\ 0.8, 0.02], min[0.2, 0.98]\}$
    $c_{21} = \{0.02, 0.2\}$
    $c_{31} = \{min[\ TC_{11}, fc_{311}], min[\ TC_{12}, fc_{312}]\}$
    $c_{31} = \{min[\ 0.8, 0.97], min[\ 0.2, 0.03]\}$
    $c_{31} = \{0.8, 0.03\}$

5. For each fault i in universe, determine the candidate membership value using $c_i = max\ \{c_{ij}\}$:

    $c_0 = 0.8$
    $c_1 = 0.2$
    $c_2 = 0.2$
    $c_3 = 0.8$

6. $FH_{T0} = \{c_o, ..., c_{m-1}\} = \{0.8, 0.2, 0.2, 0.8\}$

5.3   Combining Fault Hypotheses

Given the fault hypothesis after test 1 has been executed, FH_{T1}, and the fault hypothesis after test 2 has been executed, FH_{T2}, we would like to determine the current fault hypothesis, FH_c. The standard fuzzy set operator for intersection is the *min* operator (Klir and Yuan, 1995). Using the *min* operator, the resulting candidate membership values could be determined using:

$$c_i \in FH_c = \min\{c_i \in FH_{T1}, c_i \in FH_{T2}\}; i = 0,...,m-1$$

The primary disadvantages with this approach are that there is no rehabilitation capability for a conclusion in the model and the diagnostic granularity is low (Bearse and Lynch, to appear). The *min* operator used in the computation is the fuzzy logic equivalent to the intersection operator in set theory. The problem with this function in the diagnostic context is that once a conclusion has been reduced to a low value, it can't be rehabilitated even if a tremendous amount of additional evidence were to indicate this failure. Also, a large number of faults with the same, low truth value result from the use of the *min* operator.

In order to overcome these deficiencies, the *product* operator was selected. Using the *product* operator, the resulting candidate membership values can be determined using:

$$c_i \in FH_c = prod\{c_i \in FH_{T1}, c_i \in FH_{T2}\}; i = 0,...,m-1$$

Example. Let the current fault hypothesis be: FH_c = {0.8, 0.2, 0.2, 0.8}. This fault hypothesis was generated above by considering the candidate membership values when T0 had an outcome of pass. Now consider test T1 having an outcome of fail. If we follow the procedure described in the section above, we calculate FH_{T1} = {0.05, 0.05, 0.95, 0.95}.

Using the *product* operator, the new fault hypothesis would be computed as:

$$FH_{new} = \{0.04, 0.01, 0.19, 0.76\}$$

Note that using the *min* operator, the new fault hypothesis would have been: FH_{new} = {0.05, 0.05, 0.2, 0.8}. Even though the second fault candidate had more total evidence {0.2, 0.05} against it than the first candidate {0.8, 0.05}, the *min* operator couldn't distinguish between the two cases. The improved diagnostic granularity associated with the *product* operator makes it a better choice for this application.

5.4   Normalization

Using the *product* operator to process even a small number of tests, it is inevitable that the computation of the fault candidate degrees of truth will result in very small numbers. Since we can determine the maximum degree of truth (max. DOT), we will "normalize" the degree of truth values by dividing each membership value by the maximum degree of truth achievable.

$$\text{max DOT} = \prod_j \text{max } C_j(f_i) \tag{1}$$

where j is the test index in this example, max $DOT = 0.8 * 0.95 = 0.76$

$$FH_{norm} = \{0.053, 0.013, 0.25, 1.0\}$$

5.5   Threshold Fault Hypothesis

In general, every failure mode has some membership degree and, therefore, is in the fault hypothesis. The membership value associated with a specific failure mode is the degree of truth to the statement that the failure mode is a fault candidate. It is sometimes advantageous to weed out failure modes with low candidate membership values. This processing would reduce the number of faults in the fault hypothesis. One way to achieve this effect is to threshold each membership value according to some defined criteria. For example, if we are only interested in fault candidates that have a "normalized" truth degree of 0.1 or more, the final fault hypothesis for the example above would be F2 and F3. Likewise, if we are only interested in fault candidates that have a "normalized" truth degree of 0.5 or more, the fault hypothesis for the example would be F3. If we are only interested in fault candidates that have a "normalized" truth degree of 1.0, the fault hypothesis for the example would be F3, as well.

6.   NATURE OF CONFLICT

When processing the traditional DIMs, an inconsistency arises during the reasoning process when the fault hypothesis is determined to be the null set, ϕ. This inconsistency is referred to as conflict.

6.1   Conflict

There are three important questions that need to be addressed regarding conflict:

- What is the nature of conflict?
- Can the occurrence of conflict be avoided?
- If conflict is detected, how should it be handled?

Simpson and Sheppard believe that three fundamental reasons exist which might lead to conflict: an error occurred in testing or test recording, multiple faults exist in the system or the diagnostic model or underlying fault model is incomplete or inaccurate (Sheppard and Simpson, 1998). We consider that conflict can occur for the following reasons: modeling error (unmodeled fault, incorrect inference lists, invalid certain data assumption), incorrect test outcome (mapping of measurement to test outcome due to measurement uncertainty, noise or boundary decision), and multiple faults occur when reasoning under a single fault assumption. Assuming that test outcomes and test inferences are certain and reasoning under certainty increases the chances that conflict will occur due to the fact that during the testing process, unanticipated events will occur. Examples of conflict are discussed in the following subsections.

Unmodeled Fault. An unmodeled fault which is detectable by a test occurs. Conflict may result since the detectable fault will result in the test failing, which will then leave only a subset of the fault universe (the unmodeled fault will not be in this subset) as suspects. Subsequent testing may clear these suspects, leaving a null fault hypothesis as the only conclusion.

Incorrect Inference Lists. Assume that a fault which is detectable by a test is left off of the inference list associated with the resulting outcome for that test. Conflict may result since the detectable fault will fail this test which will then leave only a subset of the fault universe (the omitted fault will not be in this subset) as suspects. Subsequent testing may clear these suspects leaving a null fault hypothesis as the only conclusion.

Assume that a fault which is not detectable by a test is added to the inference list associated with the pass outcome. Conflict may result since the passing of test will clear the fault which then leaves only a subset of the fault universe (the added fault will not be in this subset) as suspects. Subsequent testing may clear these suspects leaving a null fault hypothesis as the only conclusion.

Certain Test Inferences. The assumption that test inferences are certain, when in fact they are not, could lead to conflict since a failure mode can be cleared when it should still be suspect. Subsequent testing may clear the remaining suspects leaving a null fault hypothesis as the only conclusion.

Certain Test Outcomes. The assumption that test outcomes are certain could lead to conflict since we are making a hard decision on which inferences to include in the fault hypothesis. This decision could eventually lead to a null fault hypothesis.

Incorrect Test Outcome. All faults implicated by an incorrect test outcome will be in the fault hypothesis. Subsequent testing may clear these suspects leaving a null fault hypothesis as the only conclusion. An incorrect test outcome can be the result of test error, measurement noise or uncertainty, or a hard decision on a test limit boundary.

Multiple Faults. Assume that the reasoner is operating under a single fault assumption and that there are two faults in independent sections of the circuit. A test that detects faults in one independent section will conflict with another test which detects the faults in the other section since the intersection of the two lists will be the null set.

6.2   Conflict Mitigation

What can be done to mitigate the occurrence of conflict? Mechanisms to minimize the occurrence of conflict are correct modeling in terms of test outcome-fault inference relationship, allowing parameters which characterize the uncertainty associated with test outcomes and test outcome-fault inferences, using the test outcome confidences during run-time by mapping measurements to more than one outcome and reasoning under a multiple fault hypothesis. Everyone should agree that we would like to create correct models such that the inferences associated with each test outcome are correct and complete. Even if the inferences associated with test outcomes in model were correct and certain, conflict could occur due to incorrect assumptions on test outcome certainty, test error or the existence of multiple faults. The AI-ESTATE standard proposes to allow the use of test outcome confidences and test inference confidences to characterize uncertain test information. Accurately capturing and using the confidences allowed in the standard are a means of reducing the occurrence of conflict. The fuzzy logic paradigm proposed in this chapter also mitigates conflict by using the test outcome confidence associated with multiple outcomes of a test at run-time in order to avoid making a hard decision which could result in conflict.

6.3   Conflict Detection and Processing

Even doing all we can to mitigate conflict, it is inevitable that conflict will occur and that the reasoning system will have to deal with it. What can be done once conflict is detected? It appears that there are two choices: make no assumptions or make some reasonable assumptions.

If conflict occurs and you do not want to make any assumptions about the model, the testing process or the reasoning process, you should stop testing and notify the user of the reasoner that a conflict has occurred. The user should also be notified that there are three possible causes: model error, test error or multiple faults.

One would have to make assumptions in order to continue processing the model and test data. If it were assumed that the model were incorrect, one wouldn't continue to reason on an incorrect model nor would one want to use the inferences in the model to determine the amount of conflict between test outcomes and their inferences. If it were assumed that the test results were in error, one couldn't account for the impact of the error on the fault hypothesis. The assumptions that the model is correct and the test results (or at least their confidences) are correct are the normal assumptions that are made in papers on system diagnosis (Sheppard and Simpson, 1998). Another usual assumption when processing inference models is that a single fault exists in the system. If conflict is detected, the reasoner could assume that the model and test results are correct and pursue a multiple fault diagnosis. Given the complexity of the test and diagnosis problem, we feel that this is a reasonable and practical approach. Other approaches have created a special conclusion called an "unanticipated result" which serves as a measure of conflict (Sheppard and Simpson, 1991a; Sheppard and Simpson, 1998).

6.4   A Conflict Detection Parameter

It would be useful to have some measure of conflict representing the degree of inconsistency of the data. In the classical approach, the fault hypothesis has a conclusion (indicating consistent data) or becomes empty (indicating a conflict). In order to expand on the binary notion of conflict, we define conflict to be in the range of 0 to 1 and compute it as follows:

$$\text{Conflict} = 1 - \frac{\text{max DOT obtained}}{\text{max DOT possible}}$$

where max DOT obtained is the maximum normalized degree of truth for any fault candidate and max DOT possible is given by (1).

If all fault candidates' DOTs are 0.0, then the Conflict will be 1.0. This situation indicates conflict exists in the traditional sense. If the max DOT obtained = max DOT possible, the Conflict = 0.0, and no conflict exists in the traditional sense. In general, the higher the test confidences, the greater the conflict parameter will be when something unexpected happens. Again, the Conflict parameter is a measure of the degree of inconsistency of the data.

7.   REHABILITATION OF FAULT HYPOTHESIS SUSPECTS

As stated, the primary disadvantage to using the *min* operator when combining fault hypotheses is that there is no rehabilitation capability for a conclusion in the model. The problem with this function in the diagnostic context is that once a conclusion has been reduced to a low value, it can't be rehabilitated even if a tremendous amount of additional evidence were to indicate this failure.

As new evidence in the form of test results is obtained, it is desirable to be able to reorder the fault candidates according to their truth value. The previous method which used the *min* operator resulted in many conclusions which had the same low truth value and couldn't be rehabilitated. The use of the *product* operator provides the desired rehabilitation capability. This operator will not raise the truth value of any conclusion, however, it will lower the truth value of other conclusions which are inconsistent with the evidence. The relative truth values will be adjusted according to how well the inference matches the evidence provided in the test outcomes.

8.   MEMBERSHIP FUNCTION DEFINITION

Let's start by assuming that a voltage to be measured must be within the following range: $5V \pm .1V$. Our test for this voltage is defined as follows:

Pass: $4.9V \le v \le 5.1V$
Fail: $v < 4.9V$ or $v > 5.1V$
where v is the voltage measured

For the moment, we will assume that the instrument tolerance is zero. We want to consider the decision making process that is used during the generation of a specific test outcome. Assuming that a measured voltage is recorded as 4.9 V, the test outcome would be pass. However, if it were 4.899 V, then the test outcome would be fail. The mapping of the test outcome is extremely sensitive to the measurement since we are forced to make a hard decision on which test outcome occurred. Intuitively, a measured voltage of 4.899V must imply some degree of "passness".

Now consider the fact that instrument tolerances are not zero. Let's assume that the DC voltage tolerance on our instrument is 25mV. Again assume that the voltage measurement is 4.9V. We are forced to decide if the test passed or failed, but in this case the true voltage is in the range 4.875V - 4.925V. This range encompasses both pass and fail test outcomes, however, we are again forced to make a "hard" decision using conventional approaches. It would be nice if we could consider the inferences associated with both test outcomes when the measured voltage is so close to the boundary.

We will construct the membership functions for the test outcomes of pass and fail and consider the results that are obtained using the measured voltage presented above. The following definitions apply to this discussion:

Nominal (N) = 5.0 V
Measurement Parameter Tolerance (T) =.1 V
Lower Limit (LL) = N - T = 4.9V
Upper Limit (UL) = N + T = 5.1V

Transition points (TP) are defined to be T/2 units greater than and less than the lower and upper limits. This choice is rather arbitrary at this point. In terms of the nominal value and the measurement tolerance, the following definitions represent the descriptions of the pass and fail membership functions. Membership function definitions could be developed using LL, UL and TP variables in the case where the test and its limits are not symmetrical.

8.1   Pass Membership Function Definition

In general, the pass membership function is defined as:

Membership Degree	Voltage Range
0	$v \le N - 3T/2$
$10(v - (N - 3T/2))$	$N - 3T/2 < v \le N - T/2$
1	$N - T/2 < v \le N + T/2$
$10((N + 3T/2) - v)$	$N + T/2 < v \le N + 3T/2$
0	$v > N + 3T/2$

For this specific example, the pass membership function is defined as:

Membership Degree	Voltage Range
0	$v \le 4.85$
$10(v - 4.85)$	$4.85 < v \le 4.95$
1	$4.95 < v \le 5.05$
$10(5.15 - v)$	$5.05 < v \le 5.15$
0	$v > 5.15$

8.2   Fail Membership Function Definition

In general, the pass membership function is defined as:

Membership Degree	Voltage Range
1	$v \le N - 3T/2$
$10((N - T/2) - v)$	$N - 3T/2 < v \le N - T/2$
0	$N - T/2 < v \le N + T/2$
$10(v - (N - T/2))$	$N + T/2 < v \le N + 3T/2$
1	$v > N + 3T/2$

For this specific example, the fail membership function is defined as:

Membership Degree	Voltage Range
1	$v \le 4.85$
$10(4.95 - v)$	$4.85 < v \le 4.95$
0	$4.95 < v \le 5.05$
$10(v - 5.05)$	$5.05 < v \le 5.15$
1	$v > 5.15$

The pass and fail membership functions are shown graphically in Figure 2.

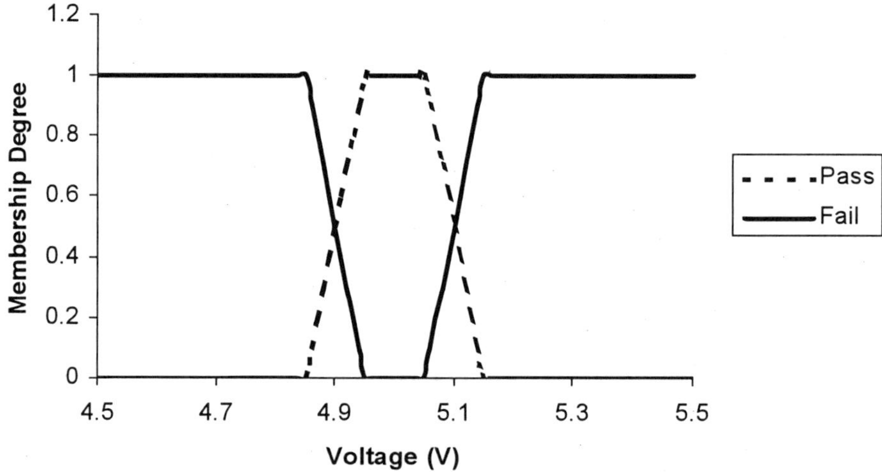

Figure 2. Membership Functions For Pass/Fail Test Outcomes

Let's reconsider a recorded voltage measurement of 4.9V along with the assumption that there is no measurement error. We can see from the equations or the graph that the membership in the fail test outcome would be 0.5 and the membership in the pass test outcome would also be 0.5.

Now consider the fact that instrument tolerances are not zero. Again, we assume that the DC voltage tolerance on our instrument is 25mV and that the voltage measurement is 4.9V. If we use the measured voltage directly, we can see from the equations or the graph that the membership in

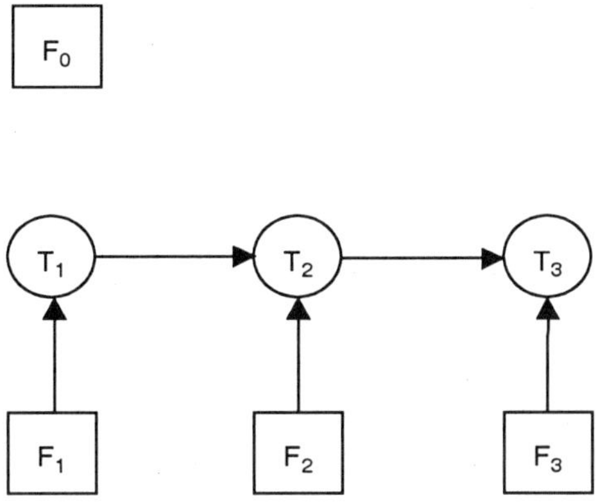

Figure 3. Simple Serial Example

the fail test outcome would be 0.5 and the membership in the pass test outcome would also be 0.5. If we considered the instrument tolerance, the true voltage is in the range 4.875V - 4.925V. The voltage of 4.875V would represent membership grades of 0.7 and 0.3 for the fail and pass outcomes respectively. The voltage of 4.925V would represent membership grades of 0.3 and 0.7 for the fail and pass outcomes respectively. One could average these two sets of membership grades to determine an assigned membership grade.

Using the membership grades from both test outcomes allows us to deal with the boundary situations effectively since we can now reason quantitatively on the inferences for both of the outcomes at the same time.

9.　　EXAMPLE

This section presents a simple example. Using the fuzzy logic methodology proposed in the chapter, we consider the four cases: certain outcomes/certain inferences, certain outcomes/uncertain inferences, uncertain outcomes /certain inferences, and uncertain outcomes/uncertain inferences.

In Figure 3, a simple serial example is shown. In this figure, F_1, F_2 and F_3 represent failure modes, F_0 represents the no fault condition and T1, T2 and T3 represent tests. The matrix associated with this graphical representation is shown in Table 3. A "1" indicates that the test outcome detects the fault. A "0" indicates that the test outcome does not detect the fault. For the moment, we are considering the case where all the test outcomes are certain.

If test T1 were to pass, one could conclude using this model that the fault hypothesis would be F_0, F_2 and F_3. Additionally, if test T1 were to pass

Table 3. Matrix Representation

	T1-fail	T1-pass	T2-fail	T2-pass	T3-fail	T3-pass
F0	0	1	0	1	0	1
F1	1	0	1	0	1	0
F2	0	1	1	0	1	0
F3	0	1	0	1	1	0

and T3 were to fail, one would conclude that the fault hypothesis would be F_2 and F_3. If we were to rerun test T3 and it were to pass, the fault hypothesis would then be the null set, ϕ. This situation illustrates conflict in a very simple way.

We wish to compare the methodology proposed in this chapter and the classical set theoretic approach. We assume that a threshold is not applied to the fault hypothesis which is to say that each fault has some membership in the fault hypothesis. Of course, this membership value may be zero. We assume that all test inferences for detections are 0.95 and that the test inference confidences associated with a specific test sum to one. All base test outcome confidences are assumed to be 0.9. These confidences could be obtained from appropriate membership functions. Table 5 represents the situation when the T1 pass outcome occurred. Table 6 represents the situation when the T1 pass outcome occurred and T3 failed. Table 7 represents the situation when the test T1 passed, test T3 failed then test T3 was re-run and passed. The degrees of truth in Tables 5-7 have not been normalized.

9.1   Certain Outcomes/Certain Inferences

We use the fuzzy logic methodology and consider the case when all of the test outcome and test inference information is certain. In the case when test outcome inferences are certain, the FL system doesn't consider the inferences associated with the "other" outcome. The results obtained are the same as the results obtained with the classical set theoretic approach. Compare rows 1 and 2 in Tables 5, 6 and 7.

9.2   Certain Outcomes/Uncertain Inferences

In this case, we assume that all test inferences are 0.95 and that the test inference confidences associated with a specific test sum to one. Outcome inferences are still certain. The certain test inference information in Table 3 is modified and presented in Table 4. Compare rows 1 and 3 in Tables 5, 6 and 7. Note that in Table 7 there is non-zero support for every conclusion even though there is "conflict" when the model is processed in the classical set theory sense. The parameterization of the test inference uncertainty has resulted in a mechanism which allows for more robust

Table 4. Uncertain Test Inference Information

	T1-fail	T1-pass	T2-fail	T2-pass	T3-fail	T3-pass
F0	0.05	0.95	0.05	0.95	0.05	0.95
F1	0.95	0.05	0.95	0.05	0.95	0.05
F2	0.05	0.95	0.95	0.05	0.95	0.05
F3	0.05	0.95	0.05	0.95	0.95	0.05

Table 5. Summary For T1-Pass

	F0	F1	F2	F3	Conflict
Classical	1	0	1	1	0
FL w/Cert. Outcomes/Cert. Inf.	1	0	1	1	0
FL w/Cert. Outcomes/Uncert. Inf. (0.95)	0.950	0.050	0.950	0.950	0
FL w/Uncert. Out. (0.9)/Cert. Inf.	0.900	0.100	0.900	0.900	0
FL w/Uncert. Out. (0.9)/Uncert. Inf. (0.95)	0.900	0.100	0.900	0.900	0

Table 6. Summary For T1-Pass And T3-Fail

	F0	F1	F2	F3	Conflict
Classical	0	0	1	1	0
FL w/Cert. Outcomes/Cert. Inf.	0	0	1	1	0
FL w/Cert. Outcomes/Uncert. Inf. (0.95)	0.048	0.048	0.903	0.903	0
FL w/Uncert. Out. (0.9)/Cert. Inf.	0.090	0.090	0.810	0.810	0
FL w/Uncert. Out. (0.9)/Uncert. Inf. (0.95)	0.090	0.090	0.810	0.810	0

Table 7. Summary For T1-Pass, T3-Fail And T3-Pass

	F0	F1	F2	F3	Conflict
Classical	0	0	0	0	1
FL w/Cert. Outcomes/Cert. Inf.	0	0	0	0	1
FL w/Cert. Outcomes/Uncert. Inf. (0.95)	0.045	0.002	0.045	0.045	0.947
FL w/Uncert. Out. (0.9)/Cert. Inf.	0.081	0.009	0.081	0.081	0.889
FL w/Uncert. Out. (0.9)/Uncert. Inf. (0.95)	0.081	0.009	0.081	0.081	0.889

processing of the model data. It is less likely that there will be conflict since the exact nature of the inference is captured with the test inference parameter.

9.3   Uncertain Outcomes/Certain Inferences

In this case, we consider the test inferences to be certain as depicted in Table 3. However, the primary test outcome is assumed to have a confidence of 0.9. The other test outcome has a test confidence of 0.1. These confidences can be determined from appropriate membership functions or provided by the user at runtime. Compare rows 1 and 4 in Tables 5, 6 and 7.

Note that in Table 7 there is non-zero support for every conclusion even though there is "conflict" when the model is processed in the classical set theory sense. The parameterization of the test outcome uncertainty has resulted in a mechanism which allows for more robust processing of the model data.

9.4   Uncertain Outcomes/Uncertain Inferences

In this case, we consider the test inferences to be uncertain as characterized in Table 4. In addition, the primary test outcome is assumed to have a confidence of 0.9. The "complimentary" test outcome has a test confidence of 0.1. These confidences can be determined from membership functions or provided by the user at runtime. Compare rows 1 and 5 in Tables 5, 6 and 7. Again, note that in Table 7 there is non-zero support for every conclusion even though there is "conflict".

Tables 5, 6 and 7 summarize the results for this simple model for the classical set theory processing and the four cases of processing using the fuzzy logic methodology. The values in these tables represent the degree of truth that the failure mode is a candidate or suspect.

10.   AI-ESTATE FRAMEWORK

The emerging IEEE AI-ESTATE set of standards is intended to provide a formal framework for exchanging diagnostic knowledge in all test environments and constructing diagnostic reasoners. To be a viable alternative in future systems, the proposed application of fuzzy logic as the primary reasoning under uncertainty mechanism for model-based diagnostics must consider the formal framework. Three areas of the AI-ESTATE standard are discussed: the Common Element Model (CEM), the Enhanced Diagnostic Inference Model (EDIM), and the Dynamic Context Model (DCM).

10.1   AI-ESTATE Common Element Model

In the section on Quantitative Model Processing, the need to determine the uncertainty associated with test outcomes is addressed. Two primary methods of providing a test outcome's base confidence were identified: determined a priori, during model development, and supplied as an element of the DIM; and derived on-line and supplied in real-time. We believe that a more robust system is provided by the later approach. Supplying the confidence in real-time will be discussed below, in AI-ESTATE Dynamic Context Model. However, a programmatic decision may be to utilize the values determined a priori.

When the base confidence is determined a priori, the *confidence* attribute of a *test*'s *outcome* entity of the AI-ESTATE CEM will contain the value to be used by the Fuzzy Logic reasoner.

10.2   AI-ESTATE Enhanced Diagnostic Inference Model

The proposed approach assumes that the test outcome-fault inference confidence values are determined a priori, during model development, and are provided as elements of the EDIM. The *confidence* attribute of the *diagnostic_inference* entities in the *outcome_inference* entity's *conjuncts* and *disjuncts* sets will contain the values to be used by the Fuzzy Logic reasoner.

10.3   AI-ESTATE Dynamic Context Model

In the proposed approach, when the confidence values associated with the test's outcomes are determined in real-time, the measured parameter(s) is transformed such that each of the test's outcomes is given a confidence value. Outcome confidence values may be provided by the user or computed using membership functions. The transformation is performed outside of the reasoner. The 2-tuples consisting of test outcome and associated confidence value for all of the test's outcomes are needed by the fuzzy logic reasoner. The reasoner requires that all of a test's outcomes and associated confidence values be available to determine the degree of truth that the failure mode is in the candidate set.

The IEEE Std 1232.2:AI-ESTATE Service Specification (1998) in ballot recirculation at the time of this writing supports the fuzzy logic reasoner requirement for providing a confidence value for each outcome of the test performed. Each *step* of a DCM *session* may have a set of *test_performed* attributes. This generalized mechanism supports performing multiple tests at the step, and supports having multiple outcomes for a single test at the step. Each element of the *test_performed* attribute set uses the *active_test* entity to assign an *actual_outcome* and an associated *actual_confidence*. The *actual_confidence* may be set using the CEM *test outcome confidence* value, or as a result of determining the value in real-time.

11.   CONCLUSIONS

The proposed methodology deals with a wide spectrum of uncertainty in a plausible manner. In particular, the methodology deals with the measurement uncertainty issue effectively since it supports gradual transitions between states (test outcomes). Ultimately, its usefulness depends on the development of appropriate membership functions. Fuzzy logic is utilized in two places in the system. The front end uses fuzzy logic to deal with the measurement uncertainty issue and the back end uses a separate fuzzy logic processor to compute membership degrees for the faults. This coupling of mechanisms has the benefits of dealing with the measurement uncertainty issue effectively as well as providing the mechanism to deal with the complete spectrum of uncertainty information allowed in AI-ESTATE. In addition, the back end processor could be replaced by a processor which deals with Dempster-Shafer theory or the MYCIN/EMYCIN calculus (certainty factors).

A case could be made for only using one test outcome for a test. However, this approach would only work well as the test outcome confidence nears 1. It is felt that, as the test confidence decreases from 1 and approaches 0.5, one must consider using the information provided by the other test outcomes. As the measurement parameter approaches the boundary between two different test outcomes, conventional approaches force one to make a hard decision on which one test outcome occurred. As such, the use of test outcome confidence for more than one outcome per test is most useful when you approach a boundary between test outcomes. When the measurement approaches the boundary, this soft mechanism provides a natural transition between states of the system. It is this region near the boundary which results in conflict and misdiagnosis when only one test outcome is considered. Reasoning from multiple outcomes for the same test reduces the chance for a conflict and is a conflict mitigation mechanism in its own right. The "hard" decision approach creates conflict due to the fact that you must decide which inference list is supported rather than providing support for inferences on both lists. The result is a robust system which is very flexible.

In general, the reasoning under uncertainty paradigm requires more model information, not less. Instead of assigning the user the additional task of generating the test outcome and test-fault inferences, one would desire a system which could let the user specify the inferences and outcomes with certainty and then perform the underlying reasoning by injecting controlled amounts of uncertainty into each outcome inference and possibly test inference. The proposed system can also be used in this manner to process the certain information provided by the user in a "reasoning under uncertainty" framework. In effect, the system uses the test outcomes and test inferences under the assumption that they are "probably" correct.

The advantages of the proposed approach are its flexibility, the processing of the test outcome inferences is commutative, increased diagnostic granularity, and the system handles measurement uncertainty intuitively. The primary disadvantages to using the *min* operator to generate the fault hypothesis is that there was no rehabilitation capability for a conclusion in the model and there is low diagnostic granularity. This lead to the analysis and selection of the *product* operator. The use of the *product* operator provided a rehabilitation capability, a measure of conflict, and an increase in diagnostic granularity.

The ability to detect conflict provides the user of the inference engine with information so that a decision on whether to continue testing can be made. The improved methodology has a method for measuring conflict which can be utilized by the user to guide the diagnostics.

12. REFERENCES

Bearse, T. and M. Lynch. "An Improved Technique for Applying Fuzzy Logic in a Model-based Diagnostics Reasoner," to appear.

Dill, H.. 1994. "Diagnostic Inference Model Error Source," *Proceedings of the IEEE Autotestcon*, pp. 391-397.

Harrison, M.C. and S. J. Henkind. 1988. "An Analysis of Four Uncertainty Calculi", IEEE Transactions on Systems, Man and Cybernetics, September/October, pp. 700-714.

IEEE Standard 1232.1-1997. *Trial Use Standard for Artificial Intelligence Exchange and Service Tie to All Test Environments (AI-ESTATE): Data and Knowledge Specification.*

IEEE Standard 1232.2-1998. 1998. *Trial Use Standard for Artificial Intelligence Exchange and Service Tie to All Test Environments (AI-ESTATE): Service Specification*, Ballot Draft.

Klir, G. and B. Yuan. 1995. *Fuzzy Sets and Fuzzy Logic:Theory and Applications*, Upper Saddle River, NJ, Prentice-Hall.

NUWCDIVNPT. 1994. *A Handbook for the Generation of Diagnostic Inference Models (Draft)*. Naval Undersea Warfare Center Division Newport.

Sheppard, J. and W. Simpson. 1991a. "Uncertainty Computations in Model-based Diagnostics," *Proceedings of the IEEE Autotestcon*, pp. 233-241.

Sheppard, J. and W. Simpson. 1991b. "A Mathematical Model for Integrated Diagnostics," *IEEE Design and Test of Computers*, Vol. 8, No.4, pp. 12-25.

Sheppard, J. and W. Simpson. 1998. "Managing Conflict in System Diagnosis," *IEEE Computers*, Vol. 31, No. 3, pp. 69-76.

Sheppard, J.. 1994. "Standardizing Diagnostic Models for System Test and Diagnosis," *Proceedings of the IEEE Autotestcon*, pp. 343-349.

Simpson, W. and J. Sheppard. 1994. *System Test and Diagnosis*, Norwell, Massachusetts, Kluwer Academic Publishers, pp. 65-89.

Chapter 4

Deriving a Diagnostic Inference Model from a Test Strategy

Timothy M. Bearse
Naval Undersea Warfare Center

Keywords: Model based diagnostics, diagnostic inference model, asymmetric diagnostic inference model, test program set, test program, test strategy, fault tree, asymmetric model generation, attribute map.

Abstract: Model based diagnostics is receiving serious consideration for use in diagnostic test programs. New weapon systems are considering model based diagnostics for its potential for life cycle cost savings and performance improvements. While the potential of model based diagnostics was targeted for use on new weapon systems, its potential for use on existing systems should not be overlooked. One issue with considering model based diagnostics for these systems is the effort to derive a useful model. It is known that the development of a model is the hard part of the problem (Dill, 1994; Simpson and Sheppard, 1994; Sheppard and Simpson, 1994). This chapter examines the feasibility of deriving a Diagnostic Inference Model from an existing fault tree. This study starts with an existing symmetric model and its resulting fault tree. From the symmetric model based fault tree, an attempt is made to rederive the original model. It will be seen that it is not feasible to rederive the original symmetric model from the fault tree. However, it will be shown that useful diagnostic information can be derived in the form of an asymmetric Diagnostic Inference Model of the system.

1.   BACKGROUND

Traditional Test Programs (TPs) utilize a static fault tree to isolate faults to the lowest level component or ambiguity group of components. The fault tree is usually documented, or may merely be embodied in the control flow of the TP's code. In either case, the conclusion reached is the result of measurements and observations taken that are used to narrow the list of suspect failures until the faulty component or ambiguity group is identified. The fault tree is static since the tests are typically designed with their position in the fault tree in mind. That is, the test is designed to consider only the list of suspected faults that remain at that point in the fault tree. Changing a test's position in a fault tree will likely subject the test to other faults. We cannot change a test's position in the fault tree without considering the effects of these faults on the test outcome. The effects of any particular failure on a test are usually poorly defined, if documented at all.

The Diagnostic Inference Model, hereafter referred to as the model, is becoming an accepted tool supporting testability analysis of systems (cards, boxes, WRAs, SRAs, etc.). Many useful Testability Figures Of Merit (TFOMs) can be derived from analysis of the model: ambiguity group sizes and composition, inherent fault isolation levels, isolation and repair distribution data, component involvement ratios, test point utilization, and more. In the context of the Integrated Diagnostics Support System (IDSS), the model is a diagnostic representation of a system that is not only used for testability analysis, but becomes a basis from which the test programs are developed. Further, the model may be used to support on-line dynamic reasoning on test outcomes to diagnose faults. When used in these ways, the model may also be used to more easily evaluate the effects of a change to a Unit Under Test (UUT) or to a UUT's TP. Thus, the model is a valuable tool throughout the life cycle of a UUT.

The model describes a system in terms of its failure modes (aspects) and the tests that detect the failure modes. These relationships may be described directly through aspect-to-test inferences or indirectly through test-to-test inferences. The inferences are usually specified and interpreted in a symmetric manner. That is, the diagnostic conclusions that may be reached based on the pass/fail results of a test are symmetric. When a test FAILs, a list of suspect failures is derived from the inferences of the model. When the same test PASSes, the same list of failures is no longer suspect, i.e., the suspects are cleared from the fault hypothesis. However, there is nothing to prevent useful diagnostic analysis and reasoning when the inferences are asymmetric in nature (Simpson and Sheppard, 1994). Diagnostic inference asymmetry results if the list of suspects when a test FAILs is different from the list of suspects to be cleared when that test PASSes.

2.   SYMMETRIC MODEL-BASED FAULT TREE

A symmetric model will be presented for a ficticious UUT. The attribute map (Simpson and Sheppard, 1994) depicting the aspect / test relationships for this model will be shown. A fault tree for diagnosing the

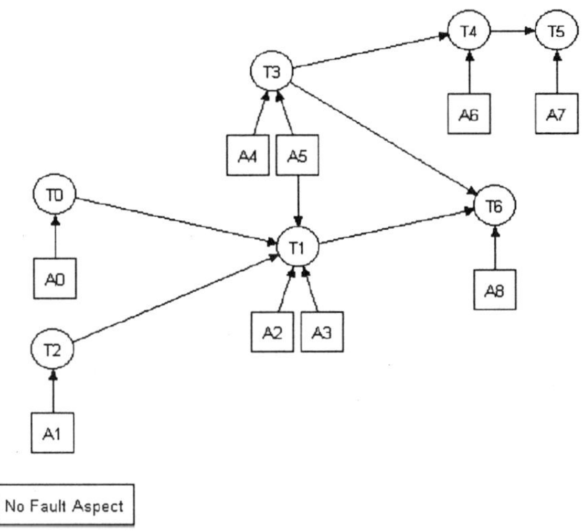

Figure 1. Diagnostic Inference Model (Graphical Representation)

UUT faults will be shown. The attribute map and the fault tree are easily obtained from a tool such as the IDSS Weapon System Testability Analyzer (WSTA) when provided with a model.

2.1   Diagnostic Inference Model

Figure 1 depicts a Diagnostic Inference Model. It depicts a graphical first order model representation (Lynch, 1995) of a symmetric inference model. As an example, from the figure, it is seen that if Test T5 FAILs, A4, A5, A6, and A7 are suspect. A7 is a direct inference. A4, A5, and A6 are transitive inferences through tests T3 and T4. If Test T5 PASSes, the same failure modes are no longer suspect. Reasoning on symmetric inferences is performed in this manner.

2.2   Attribute Map

Figure 2 depicts the attribute map for the model in Figure 1. For each failure mode (A1, etc.) in the system, the tests that will detect the presence of the failure mode are indicated (T1, etc.). Examination of the model and the attribute map shows that the two convey the same information. Testability analysis tools may statically determine the ambiguity groups by examining the fault signature for each aspect. Aspects with the same fault signature comprise the same ambiguity group. By examination of the fault signature for each failure mode in Figure 2, it is seen that each aspect's fault signature is unique, except for A2 and A3, which have the same fault signature. Thus, there is one ambiguity group of two components (assuming each failure mode represents a different component), and the remaining ambiguity groups contain one component. This information may also be determined by inspection of the model of Figure 1.

TESTS

	T0	T1	T2	T3	T4	T5	T6
N.F.	0	0	0	0	0	0	0
A0	1	1	0	0	0	0	1
A1	0	1	1	0	0	0	1
A2	0	1	0	0	0	0	1
A3	0	1	0	0	0	0	1
A4	0	0	0	1	1	1	1
A5	0	1	0	1	1	1	1
A6	0	0	0	0	1	1	0
A7	0	0	0	0	0	1	0
A8	0	0	0	0	0	0	1

ASPECTS

Figure 2. Attribute Map For Diagnostic Inference Model

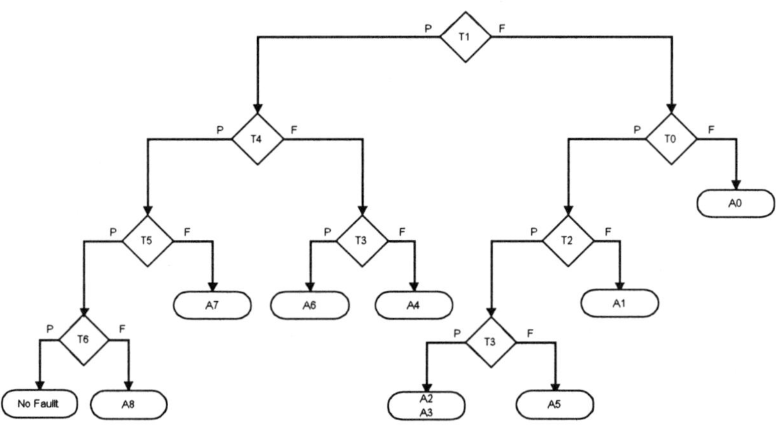

Figure 3. Fault Tree

2.3   The Fault Tree

Figure 3 depicts a fault tree (test strategy) for the system. Using the IDSS WSTA tool, the fault tree was developed while optimizing for fault isolation time. One may compare the fault tree to the graphical model to see how each failure mode is isolated. For example, a test T1 PASS outcome clears A0, A1, A2, A3, and A5 as suspects. Test T4 is then performed as a good choice to divide the remaining failure suspects. Likewise, Test T0 is a reasonable choice following a test T1 FAIL outcome.

3. DERIVATION OF AN ATTRIBUTE MAP FROM THE FAULT TREE

Diagnostic inference information is extracted from the fault tree of Figure 3. Two issues need to be addressed prior to starting the process.

First, traditional fault trees list the suspect faulty component or components at the leaf nodes of the tree. Thus, a component may show up in more than one leaf node. When a component shows up in more than one leaf node conclusion, it is necessary to flag it as being a different failure mode of the parent component. Failure to do so may result in a component being cleared as a suspect when only a particular failure mode of the component was detected by the test. In the event that it was the same failure mode that should have been in the fault tree twice, there is no adverse effect on ambiguity group sizes, merely on the number of tests required to identify the ambiguity group.

Second, in the general case, a fault tree may not have been prepared with the required test PASS / FAIL interpretations. There is an easy way to check the interpretation. To support a diagnostic inference model reasoning process, the No Fault is a possible conclusion for only the test PASS outcome. If the test go / no-go branching criteria results in the No Fault conclusion on the test FAIL leg, the test no-go leg is the test PASS outcome, and the test go leg is the test FAIL outcome.

The process begins at the leaf nodes of the tree and involves extracting information from the PASS and FAIL paths of each test. If a failure mode is listed in the FAIL path of a test, it is known that the test detects the failure mode. If a failure mode is listed in the PASS path of a test, it is known that the test does not detect the failure mode. The failure modes from both the FAIL and PASS paths of a test are the Fault Hypothesis (FH) or suspect failure modes entering the test. For example, as shown in Figure 4, Test T3 on the FAIL path of root node test T1 detects failure mode A5, but does not detect failure modes A2 and A3. The FH entering Test T3 is A2, A3, and A5. Figure 4 depicts the failure modes detected and the FH for each of the lowest level tests.

The process continues by propagating the failure modes through each test back to the root node (test) of the tree. Figure 4 shows the FH entering each test and lists the failure modes that each test detects in the tree.

The information annotated on the fault tree of Figure 4 may be represented as an attribute map, Figure 5. When a test detects a failure mode, a "1" is placed at the intersection. When a test is exposed to the failure mode via the FH and the test passes, a "0" is placed at the intersection. Note that

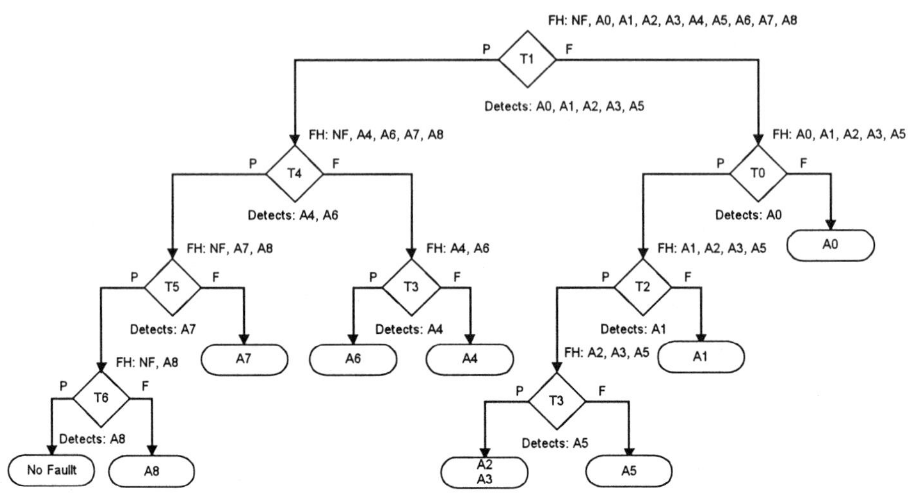

Figure 4. Annotated Fault Tree.

by definition the no fault (NF) failure mode is not detected by any test, and a "0" is placed at each intersection. After filling in all the information available from the annotated fault tree, it is seen that the attribute map is incomplete. Any map intersection left blank indicates there is no knowledge of the impact of the failure mode on the corresponding test. There is no way to complete the attribute map from the information provided by the fault tree. Due to the incomplete nature of the attribute map, symmetric inference reasoning should not be applied, as it would lead to erroneous conclusions.

TEST

	T0	T1	T2	T3	T4	T5	T6
N.F.	0	0	0	0	0	0	0
A0	1	1					
A1	0	1	1				
A2	0	1	0	0			
A3	0	1	0	0			
A4		0		1	1		
A5	0	1	0	1			
A6		0		0	1		
A7		0			0	1	
A8		0			0	0	1

ASPECT

Figure 5. Attribute Map of Information From the Fault Tree

	Test's Asymmetric Pass (Clear) List							
	T0	T1	T2	T3	T4	T5	T6	
N.F.								
A0	1	1						
A1		1	1					
A2		1						
A3		1						
A4				1	1			
A5		1		1				
A6					1			
A7						1		
A8							1	

	Test's Asymmetric Fail (Suspect) List							
	T0	T1	T2	T3	T4	T5	T6	
N.F.								
A0	1	1	1	1	1	1	1	
A1		1	1	1	1	1	1	
A2		1			1	1	1	
A3		1			1	1	1	
A4	1		1	1	1	1	1	
A5		1		1	1	1	1	
A6	1		1		1	1	1	
A7	1		1	1		1	1	
A8	1		1	1			1	

Figure 6. Asymmetric Information

4. DERIVATION OF AN ASYMMETRIC INFERENCE REPRESENTATION

An asymmetric inference representation may be derived from the incomplete attribute map in Figure 5. The asymmetric attribute map representation will have two parts. One will indicate the failure modes that may be cleared when a test PASSes. The other will indicate the failure modes that are suspect if the test FAILs.

Figure 6 depicts the asymmetric attribute mapping. It is derived from the incomplete symmetric attribute map of Figure 5 as follows. First, it is known that if the test PASSes, a failure mode marked with a "1" can not be present and may be cleared as a suspect. Therefore, for each test that detects the presence of a failure mode on Figure 5, a "1" is placed on the test's PASS (Clear) List in Figure 6. Second, it is known that if the test FAILs, the failure mode marked with a "1" is a suspect and must be considered a suspect. Therefore, for each test that detects the presence of a failure mode on Figure 5, a "1" is placed on the test's FAIL (Suspect) List in Figure 6. Third, the fact that the impact of the failure mode on a test is not known must be reflected in the carrying of the failure mode as a suspect when the test FAILs. Therefore, for each test that the intersection is blank in Figure 5, a "1" is placed on the test's Fail (Suspect) List in Figure 6. This method derives an Nth Order Asymmetric Attribute Map. That is, all aspect to test inferences are shown directly, and there are no test to test inferences.

Figure 7 depicts a graphical representation of the asymmetric diagnostic inference model. The lines with arrows represent symmetric inferences: for a test FAIL outcome the aspect is suspected, and for a test PASS outcome the aspect is cleared as a suspect. The lines with bubbles

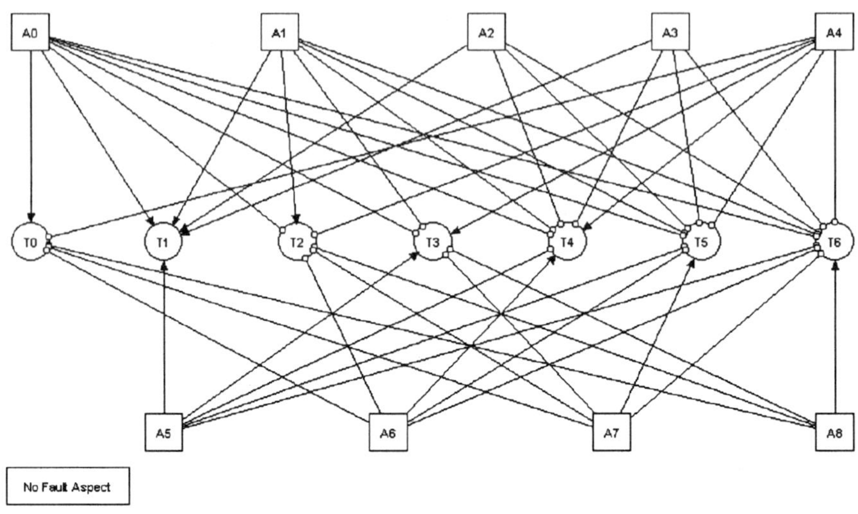

Figure 7. Asymmetric Diagnostic Inference Model (Graphical Representation)

represent negative inferences: for a test FAIL outcome the aspect is
suspected, and for a test PASS outcome the aspect is not cleared as a suspect.
The IDSS Graphical Dependency Model Editor (GRADME) prototype uses
these two inference types (symmetric and negative), since a positive inference
will never exist without a negative inference counterpart. That is, one can
never clear an aspect as a suspect on a test PASS outcome if the aspect is not
suspected on a test FAIL outcome. Figure 7 reflects the caution applied to
listing an aspect as a suspect on a test FAIL outcome when no information is
available from analysis of the fault tree.

5.   SIGNATURE ANALYSIS OF AN ASYMMETRIC DIAGNOSTIC INFERENCE MODEL

Static analysis of the symmetric attribute map of Figure 2 used the
fault signatures to statically determine the ambiguity groups. When a fault is
present, the tests that will FAIL and the tests that will PASS are
unambiguously seen in the rows of Figure 2. At this time, it will merely be
stated that the information on Figure 6 is not a fault signature. The exact
PASS / FAIL signature of the tests may not be determined from the rows of
Figure 6.

Yet, for this special case where the asymmetric information was
derived from a symmetric model-based fault tree, the information statically
depicts the ambiguity groups. Either the signatures on the PASS List or the
signatures on the FAIL List may be used to statically determine the ambiguity
groups. Aspects with the same signatures comprise the same ambiguity

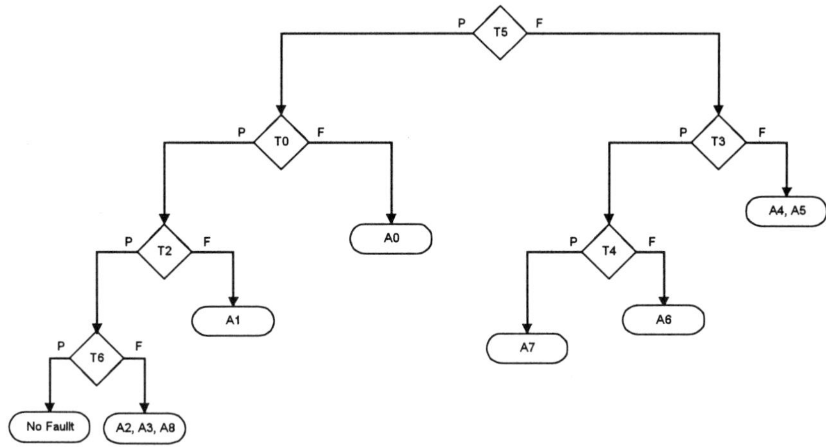

Figure 8. Fault Tree From Symmetric Model When Test T1 is Not Available

group. By examination, A2 and A3 have the same pass or fail list signature, and comprise one ambiguity group. All other failure modes have unique signatures, and are in their own ambiguity groups. The same ambiguity groups are derived from the fault signatures of Figure 2 as have been derived from the signatures in Figure 7. It will be left for a future paper to show, however, and must be emphasized here, that *in general the asymmetric attribute map test signatures do not identify ambiguity groups*. The reason it works here is that, by construction from the fault tree, for the fault hypothesis at each node in this particular fault tree, the test's inferences are symmetric.

6. ASYMMETRIC MODEL GENERATION AND ANALYSIS OBSERVATIONS

Now that we have an asymmetric model representation and have generated a new fault tree, there are several questions that are of interest. Remember, the original fault tree was generated from the symmetric, true model of the system. It reflects the actual system diagnostic behavior. We can compare the performance of the asymmetric model to that of the symmetric model. Figure 8 depicts the fault tree for the symmetric model when test T1 is not allowed. Figure 9 depicts a fault tree for the asymmetric model when test T1 is not allowed.

The first thing of notice is that the fault tree in Figure 9 is much deeper than the fault tree in Figure 8. Each test in the symmetric model has more information content than the tests in the asymmetric model. As would be expected, fewer tests are required to reach conclusions. The asymmetric fault tree shows what will be referred to as test thrashing to derive the lowest

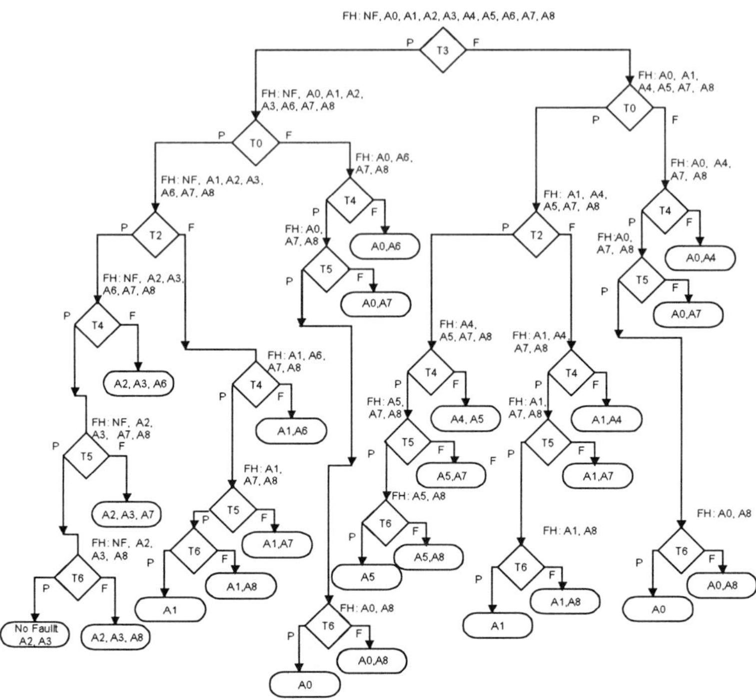

Figure 9. Fault Tree From Asymmetric Model When Test T1 is Not Available

level ambiguity groups. With test T1 not allowed (the worst case), there are no tests available that cleanly split the fault hypotheses, and therefore combinations of tests are necessary to complete the isolation process. In many cases, the lowest level ambiguity groups require information from executing all available tests before no further information can be applied to the solution.

Figure 10 shows a comparison of the performance of the two trees for each failure mode. First, it is important to note that in all cases the ambiguity groups for the asymmetric model correctly contain the faulty component. In fact, many of the ambiguity groups are the same between the two models.

Second, there are paths in the asymmetric model tree that will never be followed when compared to the actual diagnostic behavior of the tests. However, there is no way to know which paths will not be followed based on the information in the asymmetric model. This is why, even though the ambiguity groups would lead one to believe isolation to A0 and A1 would be achievable, the actual diagnostic behavior of the tests will not allow achievement of that level of isolation. From Figure 10, the average number of tests required to isolate the faulty component in the symmetric model is 3.1

tests, and the average number of tests required to isolate the faulty component in the asymmetric model is 5.2 tests. From the fault trees, the average number of tests required to isolate to the faulty ambiguity groups in the symmetric model is 3.0 tests, and the average number of tests required to isolate to the ambiguity groups in the asymmetric model is 4.95 tests.

Failure Present	Symmetric Model Test Pass/Fail Isolation Sequence	Symmetric Model Ambiguity Group	Asymmetric Model Test Pass/Fail Isolation Sequence	Asymmetric Model Ambiguity Group
NF	T5(P), T0(P), T2(P), T6(P)	{NF}	T3(P), T0(P), T2(P), T4(P), T5(P), T6(P)	{NF,A2, A3}
A0	T5(P), T0(F)	{A0}	T3(P), T0(F), T4(P), T5(P), T6(F)	{A0,A8}
A1	T5(P), T0(P), T2(F)	{A1}	T3(P), T0(P), T2(F), T4(P), T5(P), T6(F)	{A1,A8}
A2	T5(P), T0(P), T2(P), T6(F)	{A2, A3, A8}	T3(P), T0(P), T2(P), T4(P), T5(P), T6(F)	{A2, A3, A8}
A3	T5(P), T0(P), T2(P), T6(F)	{A2, A3, A8}	T3(P), T0(P), T2(P), T4(P), T5(P), T6(F)	{A2, A3, A8}
A4	T5(F), T3(F)	{A4, A5}	T3(F), T0(P), T2(P), T4(F)	{A4, A5}
A5	T5(F), T3(F)	{A4, A5}	T3(F), T0(P), T2(P), T4(F)	{A4, A5}
A6	T5(F), T3(P), T4(F)	{A6}	T3(P), T0(P), T2(P), T4(F)	{A2, A3, A6}
A7	T5(F), T3(P), T4(P)	{A7}	T3(P), T0(P), T2(P), T4(P), T5(F)	{A2, A3, A7}
A8	T5(P), T0(P), T2(P), T6(F)	{A2, A3, A8}	T3(P), T0(P), T2(P), T4(P), T5(P), T6(F)	{A2, A3, A8}

Figure 10. Comparison of Fault Tree Isolation Sequences

Whichever way you choose to look at the result, the symmetric model outperforms the asymmetric model. It is the penalty that must be paid when asymmetry is introduced.

Third, based on the actual system performance, there are only a subset of ambiguity groups that will ever be reached. Again, there is no way to know which ambiguity groups will not be reached based on the information in the asymmetric model.

In summary, when an asymmetric model is generated from a fault tree, the actual system diagnostic performance will not necessarily agree with the system performance predicted by analysis of the asymmetric model when tests are not allowed. It depends on which test is not allowed. The most that is known is that the actual performance may be better than the predicted performance. In the general case, where the tests are truly asymmetric in their inferences, the optimum tree is generated and the system diagnostic performance TFOM predictions should be accurate.

7.   CONCLUSIONS

It was seen that there is no way to complete a symmetric attribute map from the information provided by a fault tree. Thus, symmetric inference reasoning should not be applied. However, it was shown that an asymmetric representation may be obtained.

A concern was noted for fault trees generated from the asymmetric model. When key tests were marked as Not Allowed, the fault tree that was generated was somewhat inefficient when compared to the test sequences followed by marking the same tests as not available in the symmetric model. This is the result of how the asymmetric model was developed; when no knowledge of the impact of an aspect on a test was known, it was indicated that the aspect could cause the test to FAIL, but could not be cleared on a PASS. This level of caution results in more work (tests) being necessary to obtain the lowest level ambiguity groups. Individual tests provide small gains in information with regard to their asymmetric coverage. Combinations of tests were needed to extract all the available information. The ambiguity always contained the faulty aspect when the true diagnostic behavior of the tests were used. This behavior suggests that when an asymmetric model is derived from a fault tree, analysis of the model in the area of the asymmetric inferences should be performed to remove as many of the asymmetric inferences as possible. The ambiguity groups will always identify the faulty component, but smaller ambiguity groups and shorter test sequences will be the reward for this effort.

Typically, the task of fully understanding the behavior a test for an existing system is, to a large extent, a re-engineering process. Experience has taught us that such re-engineering is not a trivial task. Any information available on the system will make the task easier. This information may include the theory of operation, functional descriptions, performance and design requirements data, requirements allocation data, schematics, parts lists, test requirements data, failure data, and other information. In the context of the technique described herein for generating a diagnostic inference model from a test strategy, all this information can be analyzed and used to embellish the diagnostic model. The potential added value of this process is that there is an automated technique for identifying the areas in which to focus. Cost, schedule and performance advantages over manual approaches are anticipated.

8. REFERENCES

Dill, H.H.. 1994. "Diagnostic Inference Model Error Sources," Proceedings of the IEEE AUTOTESTCON, New York: IEEE Press, pp. 391-397.

Lynch, M.L., 1995. "Diagnostic Inference Model Representations," Integrated Diagnostics Support System Technical Note, Naval Undersea Warfare Center, Newport, RI

Sheppard, J.W., and W.R. Simpson. 1994. "Dependency Modeling Pitfalls," *IEEE AUTOTESTCON'94 Conference Record*, New York: IEEE Press, pp. 717-720.

Simpson, W.R. and J.W. Sheppard. 1994. *System Test and Diagnostic*, Norwell, Massachusetts: Kluwer Academic Publishers.

Chapter 5

Inducing Diagnostic Inference Models from Case Data

John W. Sheppard
ARINC Incorporated

Keywords: Diagnostic inference models, case based reasoning, diagnostics, system test, Dempster-Shafer, nearest neighbor.

Abstract: Recent attention to using "case-based" reasoning for intelligent fault diagnosis has led to the development of very large, complex databases of diagnostic cases. The performance of case-based reasoners is dependent upon the size of the case base such that as case bases increase in size, it is usually reasonable to expect accuracy to improve but computational performance to degrade. Given one of these large case bases, it is advantageous to attempt to induce structure from the case base whereby the diagnostic process can be made more efficient. In addition, certainty properties of case based reasoning make fault diagnosis difficult, in which case inducing structure and applying a method for reasoning under uncertainty becomes advantageous. In this chapter, we discuss an approach to analyzing a diagnostic case base and inducing a compact knowledge base using the diagnostic inference model with which efficient diagnostics can be performed. We then apply an approach to reasoning using Dempster-Shafer theory to improve the diagnostics further with the resultant model.

1. INTRODUCTION

In the process of developing an optimized approach to system test and diagnosis, it is desirable to apply as much available information to the task as possible. Test and diagnosis is a process of information gathering and analysis, and several approaches have been applied to the problem. Techniques classified as "advanced" frequently fall in the realm of artificial intelligence and can include rule-based methods, model-based methods, and instance-based methods.

Rule-based methods reason with sets of heuristics (i.e., rules of thumb) to draw inferences from test results to make a diagnosis. Sometimes these rules include information on ways to "optimize" the test process. This approach proceeds from the premise that the rules correspond to a reasoning process that represents approaches "experts" apply to problem solving. Model-based methods use a model of the system to be tested to determine "optimal" approaches to testing. Diagnosis involves reasoning about this model under the premise that anything that can be concluded about the model would also be reflected in the actual system. Instance-based methods treat the test problem "extensionally." In other words, rather than capturing rules for inference or modeling the system to be tested, instance-based methods tackle the problem by storing examples of past experience. The premise associated with this approach is that most problem solving involves considering and adapting past experiences rather than using some underlying domain theory or reasoning process.

Diagnosis is a favorite problem for researchers in artificial intelligence and operations research because of the difficulty of the problem, the generality of the problem, and the applicability of a variety of techniques to solving the problem. As one might expect, no single approach can be claimed as the "best" approach to diagnosis. For this reason, it makes sense to consider combining approaches, taking advantage of the best features of each of the individual techniques to yield a robust, "multi-strategy" approach to solving the problem.

In this chapter, we discuss ideas for combining an instance-based method (case based reasoning) with a model-based method (diagnostic inference modeling) to improve the robustness of diagnostic models and diagnostic problem solving. The primary objective of this exploration is to consider an approach to constructing diagnostic inference models from past experience to produce robust models of the system to be tested. The advantages of such a combination include robust knowledge (as reflected in the stored experiences), increased efficiency (resulting from analysis of the generated model), and improved diagnostic accuracy.

2.   DIAGNOSIS WITH CASE BASED REASONING

Case based reasoning (CBR) is a method of reasoning that combines elements of instance-based learning and data base query processing. Typically, CBR systems consist of the following elements:

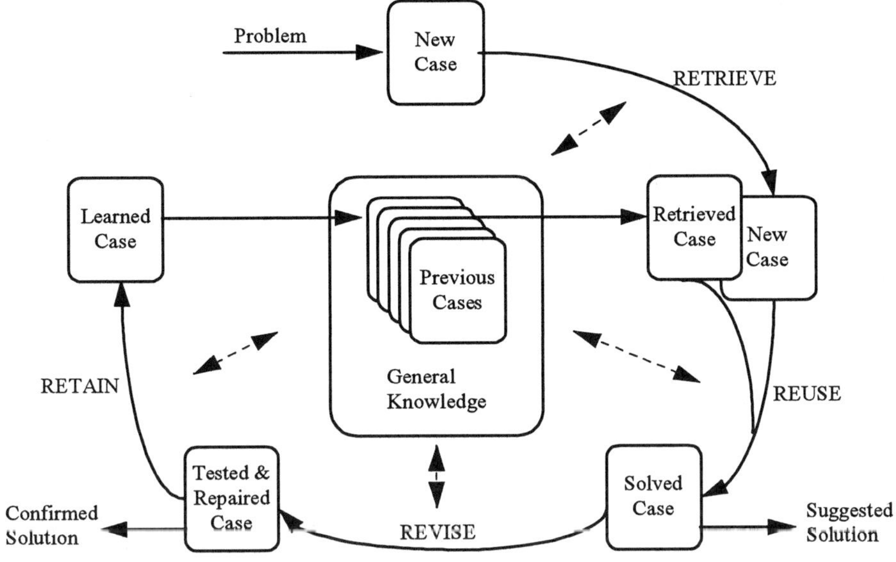

Figure 1. CBR Cycle (Aamodt and Plaza, 1994)

1. A case base
2. A case retrieval system
3. A case adaptation system
4. A case modification system
5. A case update system

These systems work together in what is called the "CBR" cycle. This cycle is illustrated in Figure 1 and consists of four activities—case retrieval, case reuse, case revision, and case retention.

When a problem is presented to the CBR system, it is treated as a "new case." This case is matched with other cases in the case base (also referred to as "memory") until the case most similar to the new case is found. The process of identifying the most similar case is referred to as *retrieval* and may use general knowledge about the problem to guide the search. Once a case is retrieved from the case base, it is reused by the CBR system to recommend a solution to the problem. The process of *reuse* applies general knowledge about the problem to adapt the solution associated with the retrieved case to the new problem. The resulting solution is then recommended to the user. Ultimately, a solution will be found for the problem at hand. This solution may be different from the recommended solution. At this point, the solution associated with the case is further *revised* to account for the actual solution. This revision may include adding, deleting, or modifying information in the case to facilitate proper application of the case in the future. Finally, the new case (with the revised solution) is *retained* in the case base for future application.

Test and diagnosis can use CBR in several ways. The simplest method involves defining a case as a collection of test results and attempting to determine an appropriate diagnosis given these results. This use of CBR is equivalent to a classification problem as characterized by instance-based learning (IBL). When using the IBL approach, the CBR cycle is simplified in that no adaptation during reuse is required. Further, revision only occurs when the diagnosis was wrong. At that point, only the diagnosis is changed in the case. Finally, the retrieval process is very simple. All of the cases are nothing more that feature vectors with an associated diagnosis $\langle f_1, f_2, \ldots, f_n; d_i \rangle$. The features in the feature vector correspond to test results and may be unknown. Retrieval then consists of "matching" the new case with all of the cases stored in the case base and selecting the most similar case.

When considering possible similarity metrics, numerical features are frequently compared using a member of the family of L_p norms. An L_p norm is defined to be

$$L_p(\mathbf{x}', \mathbf{x}'') = \left(\sum_i (x_i' - x_i'')^p \right)^{1/p}.$$

The most common values for p are 1, 2, and ∞ and yield "Manhattan" distance, Euclidean distance, and max-norm distance respectively. Specifically, these metrics can be computed as:

$$L_1(\mathbf{x}', \mathbf{x}'') = \sum_i (x_i' - x_i'')$$

$$L_2(\mathbf{x}', \mathbf{x}'') = \sqrt{\sum_i (x_i' - x_i'')^2}$$

$$L_\infty(\mathbf{x}', \mathbf{x}'') = \sum_i \max\{x_i', x_i''\}$$

If we were using pass/fail results for testing, we would use either L_1 or L_2 (which would be equivalent). Note this is exactly what is done with fault dictionary-based diagnosis. With real values, we would most likely use L_2. Symbolic results are a bit more complicated and would require something like Stanfill and Waltz's "value difference metric" (1986). Regardless of the metric, retrieval would be done as

$$\mathbf{case} = \underset{\forall c \in \text{CASE\_BASE}}{\arg \min} \{\delta(\mathbf{new}, \mathbf{c})\}.$$

A common variant to this "nearest-neighbor" method is called "*k*-nearest neighbor" and involves retrieving the *k* nearest cases from the case base and combining (or voting among) the recommended diagnoses. This approach has been demonstrated to approach the Baye's optimal solution as *k* increases. Other variants involve weighting the solutions based on distance and weighting the features based on significance (or some other factor).

Note that nearest neighbor classification assumes that the mathematical "space" corresponding to the classification problem is a *metric space*. A metric space is defined to be a pair $\langle \Sigma, \delta \rangle$ consisting of a set of points Σ and a distance measure δ. Further, this distance measure is defined to be single-valued, non-negative, and real-valued for all members of the set Σ, i.e., $\forall s_i, s_j \in \Sigma, \delta(s_i, s_j) \in \mathfrak{R}^+ \cup \{0\}$ with the following properties:

1. Reflexivity: $\delta(s_i, s_j) = 0$ *iff* $s_i = s_j$,
2. Symmetry: $\delta(s_i, s_j) = \delta(s_j, s_i)$,
3. Triangle Inequality: $\delta(s_i, s_k) \le \delta(s_i, s_j) + \delta(s_j, s_k)$.

This assumption will be important when we assess the expected performance of case-based methods in fault diagnosis.

A second approach to applying CBR to diagnosis considers a complete "plan" for diagnosis as a case. This approach assumes the plans can be determined in connection with different situations and that the situations can be characterized prior to diagnosis. The case features would be this characterization, and the case solution would be the recommended plan. Due to the complexities of developing these plans and characterizing the context of diagnosis, this approach is not usually used.

3. FAULT DICTIONARIES AS CASE BASES

One common approach to fault isolation related to case-based reasoning is based on the fault dictionary. Fault dictionaries define a mapping from combinations of input vectors and output vectors to faults. Formally, this is represented as $FD: I \times O \to F$ where FD is the fault dictionary, I is the space of input vectors, O is the space of output vectors, and F is the space of faults. At a more basic level, this can be represented as $FD: \{0,1\}^n \times \{0,1\}^m \to F$. This representation makes the fact the vectors are binary explicit. In the simplest case, diagnosis can be performed with a fault dictionary by finding a direct match between the input/output vectors and a fault in the dictionary. Indeed, with a proper model, high confidence tests, and a reasonable fault universe, many faults will be identified in this manner.

For illustration purposes, we will use a simple digital circuit (Abramovici, Breuer, and Friedman, 1990). This circuit is given in Figure 2. From this figure and assuming a single-stuck-at fault model, we can identify 26 possible stuck-at faults. Each stuck-at fault is denoted as x_i where x is a letter matching the line where the fault occurs, and i is either 0 or 1 (denoting stuck-at-0 or stuck-at-1 respectively). We close the fault universe by defining a special "fault" in which no fault has been detected and denote this *nf*. The fault dictionary would then include the input vectors (i.e., the patterns applied to lines a, b, and c) and the expected response vector (in this case the value at m). Also associated with that entry would be the list of faults detected should the response be in error.

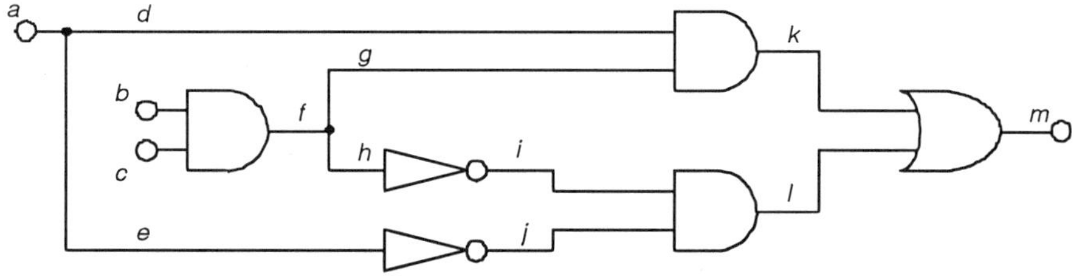

Figure 2. Sample combinational circuit.

In the following sub-sections, we will discuss how to use this information to construct the fault dictionary and how to diagnose faults with the fault dictionary. We will then discuss the problem of diagnosis given inexact matches with elements in the dictionary and describe the nearest neighbor approach for diagnosis with inexact matches.

3.1   Constructing a Fault Dictionary

To explain how to construct a fault dictionary, we will walk through the building of the fault dictionary for the circuit given in Figure 2. In this circuit, we see that there are only three input lines, therefore there are only eight possible input vectors (disregarding timing and faults other than stuck-ats). For our example, we can examine all eight inputs; however, in general, enumerating all possible vectors would be too costly. If the circuit had been a sequential circuit, then the three input lines might require several additional tests because of the sensitivity of the circuit on the previous state of the circuit. Not surprisingly, this combinatorial explosion is significantly worse for larger circuits. Several tools such as LASAR (Richman and Bowden, 1985; Grant, 1986) provide assistance to the modeler in developing input vectors and detecting stuck-ats and other faults at output vectors.

Table 1. Ambiguity groups for sample circuit.

Number	Ambiguity Group	Number	Ambiguity Group
1	a_0	8	g_1
2	a_1	9	i_0, h_1, l_0, j_0, e_1
3	b_1	10	i_1, h_0
4	c_1	11	j_1, e_0
5	d_1	12	k_0, d_0, g_0
6	f_0, b_0, c_0	13	k_1, l_1, m_1
7	f_1	14	m_0

Limiting ourselves to the combinational case (and the example in Figure 2), we begin constructing the fault dictionary by considering the possible input patterns. Each input pattern can be regarded as a test. For example, one test might be the pattern (0 1 1). Tracing through the circuit, we would expect the output of the circuit to be (0). If the value is (1) then a fault must be present in the circuit. The question then becomes, what failure modes (i.e., stuck-at faults) can cause the erroneous output? Again, examining the circuit identifies a_1, b_0, c_0, d_1, f_0, i_1, h_0, k_1, l_1, or m_1 as possible causes. Similarly with the pattern (0 0 0), we would expect the output of the circuit to be (1) and an output of (0) will implicate one of the following faults: a_1, b_0, b_1, c_0, c_1, d_1, f_0, f_1, h_0, i_1, j_1, k_0, k_1, l_1, or m_1.

Completing the analysis finds several failure modes are "ambiguous," meaning no test vectors can differentiate them. These ambiguity groups, which were taken from (Abramovici *et al.*, 1990) are shown in Table 1. One approach used for determining ambiguous faults is called "fault collapsing" and consists of identifying lines in the circuit that will have identical values regardless of input or fault because of the logical nature of the gates in the circuit. For example, if we examine the initial AND gate with inputs b and c, we note that any of b_0, c_0, and f_0 must be indistinguishable because either b_0 or c_0 (or both) will force f to have a value of zero, whether or not f is faulty. This approach to ambiguity analysis is incomplete, as we will show later, but provides a first cut on the ambiguity in the fault dictionary. From this point forward, we will refer to a particular ambiguity group by the first member of the group.

As mentioned above, the fault dictionary can be constructed in one of two ways. The first approach includes an entry for all of the input/output vectors, and each entry includes the list of faults detected.[1] For the circuit in

---

[1] In an attempt to improve the robustness of the standard fault dictionary (based on the single stuck-at fault model), Richman and Bowden (1985) introduced the concept of a "possible detection" in their modern fault dictionary. A possible detection is one that

Figure 2, this means 16 entries would be required; however, we can eliminate all of the "nominal" cases (i.e., the cases where the outputs are correct) since the only associated fault would be nf.[2] This reduces the number of entries in the dictionary to eight. For example, entries to the table for the example vectors would be

$$0\ 1\ 1\ 1 : a_1\ d_1\ f_0\ i_1\ k_1.$$
$$0\ 0\ 0\ 0 : a_1\ b_1\ c_1\ d_1\ f_0\,f_1\ i_1\ j_1\ k_0\ k_1.$$

Diagnosis using this form of fault dictionary consists of collecting the sets of detections corresponding to erroneous outputs and taking the intersection of these sets. The fault or faults reported would be given by this intersecting set. For example, suppose in addition to patterns (0 1 1) and (0 0 0), we also ran pattern (1 1 1). The entry in the fault dictionary for this pattern would be

$$1\ 1\ 1\ 0 : a_0\ f_0\ k_0\ m_0.$$

Assuming both of the vectors (0 1 1) and (1 1 1) yielded erroneous outputs, we would take the intersection of the two sets which would lead to a diagnosis of f_0 (intersection of the set $\{a_1\ d_1\ f_0\ i_1\ k_1\}$ and $\{a_0\ f_0\ k_0\ m_0\}$). Note that if we had a circuit with more than one output, we could construct separate fault dictionaries for each output, "sensitizing" the appropriate paths leading to that output to determine the faults detected. These dictionaries could be combined into one large dictionary, but this may lead to further confusion should there be a mismatch between the target signature and the signatures in the dictionary.

The second approach to representing the fault dictionary is to construct a table in which each input vector corresponds to a row in the table. The columns of the table correspond to all of the ambiguity groups in the circuit. Each cell in the table contains the expected output from the circuit. The fault dictionary for our example circuit would be represented in this form as in Table 2. This dictionary assumes eight tests as follows:

$t_1 : 0\ 1\ 1$	$t_2 : 1\ 1\ 0$	$t_3 : 1\ 0\ 1$	$t_4 : 1\ 1\ 1$
$t_5 : 0\ 0\ 1$	$t_6 : 0\ 0\ 0$	$t_7 : 0\ 1\ 0$	$t_8 : 1\ 0\ 0$

"may" be detected when an unexpected test result is encountered. Possible detections arise when indeterminate states propagate through the system.

[2] Unless there are undetected failure modes in the circuit. Of course, in this case, these nondetections would be ambiguous with nf, and we decided to refer to the ambiguity group by a representative member, namely nf. Thus the reduction is still valid.

Table 2. Fault dictionary for sample circuit.

	a_0	a_1	b_1	c_1	d_1	f_0	f_1	g_1	i_0	i_1	j_1	k_0	k_1	m_0	nf
t_1	0	1	0	0	1	1	0	0	0	1	0	0	1	0	0
t_2	1	0	0	1	0	0	1	1	0	0	1	0	1	0	0
t_3	1	0	1	0	0	0	1	1	0	0	1	0	1	0	0
t_4	0	1	1	1	1	0	1	1	1	1	1	0	1	0	1
t_5	1	0	0	1	1	1	0	1	0	1	1	1	1	0	1
t_6	1	0	1	1	1	1	0	1	0	1	1	1	1	0	1
t_7	1	0	1	0	1	1	0	1	0	1	1	1	1	0	1
t_8	1	0	0	0	0	0	1	1	0	0	1	0	1	0	0

Diagnosis using the second format matches the results of running the tests with the columns in the table. For example, suppose we run all eight tests and get (1 0 0 1 1 1 1 0) as the set of responses. This pattern would match both d_1 and i_1, indicating ambiguity between the two associated groups.[3] The significant observation to be made here is that ambiguity is determined by the actual tests used to test the circuit, and selecting a subset of possible test vectors could result in a variety of different ambiguity groups. For example, if we only evaluated t_1, t_2, t_3, and t_4, we would find a_1 is now ambiguous with both d_1 and i_1.

3.2 Diagnosis with Nearest Neighbor Classification

The most common approach to generating and using fault dictionaries for diagnosis is based on the instance-based or case-based approach discussed in the previous section. We will show later in this chapter how the model-based approach can be used to generate more robust diagnostics. For now, we will consider the approach of matching vectors of test results. In particular, we want to consider the case where the vector of test results fails to match any of the columns in the fault dictionary.

Given the single stuck-at fault model, we assume the circuit simulation accurately reflects the performance of the actual circuit. In other words, we assume the only faults of interest to us are stuck-at faults, these faults are accurately represented in the circuit model, and only one of these faults will be encountered at a time. Given these assumptions and the fact digital circuit models are deterministic (i.e., the outputs are directly determined by the inputs and, in the case of sequential circuits, the internal state), whenever the fault signature fails to match any exemplars in the fault

---

[3] The previous set of ambiguities was constructed assuming all of the test vectors were available. This analysis was performed using a fault-collapsing technique (Abramovici *et al.* 1990). However, as we see, d_1 and i_1 are also inherently ambiguous, as are g_1 and j_1, illustrating that fault collapsing is not sufficient for robust ambiguity analysis.

dictionary, the circuit must be exhibiting behavior that was not represented in the circuit model (i.e., the problem lies in the fault model, not the test results).

Debaney and Unkle (1995) assert, "In practice, it is very seldom that an observed fault signature has an exact match in the fault dictionary." This result motivates the need for an "inexact" pattern matching algorithm when using a fault dictionary. Inexact matches may be caused either by submitting noisy inputs to the circuit, introducing noise in the output, omitting an important signature from the dictionary, or having to deal with multiple faults or indeterminate states, thus invalidating the signatures. Noisy input data may yield unexpected results, but since we have control over the inputs, noise at this level indicates a problem in the test equipment or the test itself. Noisy output data indicates either a problem in the test equipment, a failure mode not included in the model, or an error in the circuit model. It is possible that vectors have not been included in the fault dictionary which, when applied to the circuit, would yield unexpected results, but having control over the test process should preclude applying those input vectors.

The current practice for processing inexact matches in fault dictionaries applies various distance measures to find the column in the dictionary that most closely matches the target vector. The literature reports two different distance measures used with the fault dictionary:

1. Hamming distance: $\delta(s_i, s_j) = \sum_b \left| s_i^b - s_j^b \right|$, where b denotes the bit compared,

2. Overlap metric: $\delta(s_i, s_j) = \dfrac{\left| s_i \cap s_j \right|}{\left| s_i \cup s_j \right|}$.

Given these distance measures, the fault dictionary can be characterized as a metric space. However, we still need to determined whether any distance measure satisfying the above properties is reasonable to apply to our problem.

As mentioned above, the presence of possible detections complicates the matching algorithm in that matches are considered but mismatches are ignored. For example, the LASAR approach uses a variation on nearest-neighbor matching in which it is assumed all of the vectors initially match the target vector, and scores for signatures in the fault dictionary are penalized when a mismatch occurs. With mismatches on definite detections and non-detections, a large penalty is applied, but when the mismatch occurs on a possible detection, only a small penalty is applied. This has the effect of "weighting" the test attribute based on the certainty of the test being able to detect (or clear) a particular fault.

4. CASE BASE COMPRESSION

Generating and using full fault dictionaries or case bases for complex systems, in general, is not practical due to the intense computational requirements, even when limited to simplified fault models. For this reason, several approaches exist to reduce the size of the case base by either not generating entries that would add little diagnostic information or by compressing the case base following generation (Tulloss, 1978; Tulloss, 1980; Ryan, 1994).

The idea behind most of the case-base compression techniques is to apply a greedy heuristic to decide whether or not to include a case in the case base. The greedy approach is used since the problem of determining the smallest set of examples to isolate all faults is reducible to the set-covering problem which is known to be *NP*-complete, In general, a heuristic evaluation function (HEF) is used (e.g., number of new faults detected or isolated), and cases are added that improve the value of the HEF the "most." Alternative approaches begin with all of the cases or signatures in the case base and "drop" faults with HEFs with the lowest value.

These compression techniques are analogous to the problem of selecting relevant features in pattern recognition. The two most common approaches to feature selection are step-wise forward selection and step-wise backward selection (Devijver and Kittler, 1982). In step-wise forward selection, features are added until classification accuracy begins to degrade, and the features are selected incrementally based on maximizing classification accuracy. In step-wise backward selection, features are removed using the same criteria. These approaches are analogous to determining which tests to include in the fault dictionary and form a method of compressing the dictionary along the test axis.

In the pattern recognition literature (Dasarathy, 1991), approaches to reducing the size of the case base that more closely relate to the common fault dictionary compression methods are known as editing methods. Frequently, it has been found that editing can improve performance of nearest neighbor classification, especially when the case base has a large number of noisy features or noisy examples. Early work by Wilson (1972) showed that examples could be removed from a set used for classification and suggested that simply editing would frequently improve classification accuracy (in the same way that pruning improves decision trees (Mingers, 1989)). Wilson's algorithm classifies each example in a data set with its own k nearest neighbors. Those points that are incorrectly classified are deleted from the case base, the idea being that such points probably represent noise. Tomek (1976) modified this approach by taking a sample (> 1) of the data and classifying the sample with the remaining examples. Editing then proceeds using Wilson's approach. These approaches are analogous to determining which fault signatures to include in the fault dictionary and form methods of compressing the dictionary along the fault axis. Any compression method discards information. In the most benign case, it discards "unneeded" information resulting in no effective loss in information. However, when faced with the possibility of being unable to match signatures exactly, it is unclear what information is unneeded, and compression may aggravate the matching problem.

5.   PROBLEMS WITH CASE BASED DIAGNOSIS

For maintenance to be cost effective, the troubleshooting strategy applied to a unit under test must be efficient and effective. Efficiency is important for reducing the cost of doing maintenance by optimizing the required maintenance resources. Effectiveness is important since ineffective maintenance leads to increased logistics costs and sparing requirements. Effective includes the important attribute of accuracy. Ineffective repair and its effects can be attributed directly to lack of effective troubleshooting.

We claim that applying the nearest neighbor classification method to outcome-based diagnosis such as that used with most case-based reasoners and fault dictionaries leads to ineffective diagnostics and, thereby, ineffective repair. In fact, we find in (Abramovici *et al.*, 1990) that the following problems are already known to exist in using fault dictionaries. First, the computational requirements for computing fault dictionaries is quite high, making generating a dictionary for a large circuit with long test sequences expensive. Thus dictionary compression is required. Second, fault dictionaries depend on a pre-defined "fault universe" and "test universe." In other words, the fault dictionary will find only the faults specified in the dictionary, and these faults can be found only with the specified set of tests. The primary assumption here is that the fault is a member of the set defined in the fault dictionary and the output vector of tests is in error in one or more bits. Nearest neighbor would treat this like a noisy signal problem, finding the existing candidate with the closest match of attributes. Finally, Abramovici *et al.* (1990) note that the nearest-neighbor approach to matching inexact patterns in the dictionary, while effective in many cases, "is not guaranteed to produce the correct diagnosis."

In the following sections, we will focus on this third issue and illustrate that, in fact, nearest neighbor is a poor choice for handling inexact matches in the general case. We will discuss sources of error in the nearest neighbor approach and suggest that, due to the discrete nature of the problem, nearest neighbor is less appropriate than other available approaches for diagnosis with the same data. Further, we will show that the primary cause of

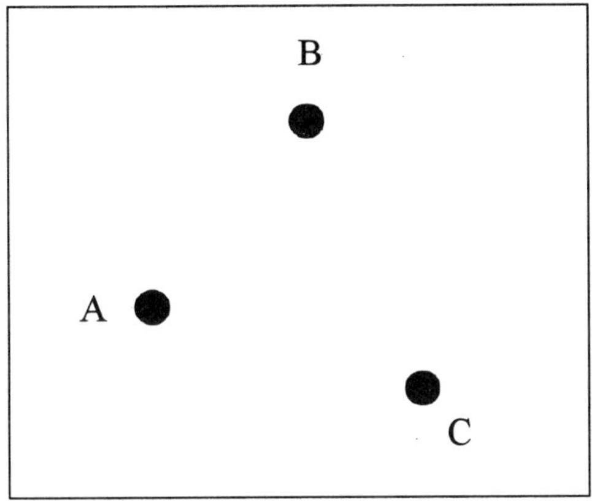

Figure 3. Nearest neighbor exemplars.

nearest neighbor's poor performance is its focus on the classification (i.e., fault) space, rather than the attribute (i.e., test) space.

5.1   Sources of Error Using Nearest Neighbor

One of the simplest ways to understand how nearest neighbor classification works is by casting the approach in a geometric framework. In general, we can think of points stored in a data base as representing concepts to be identified. When presented with a new point, we look at the "exemplars" stored in the data base to help us decide which concept best classifies the new point. In a sense, we are looking for a dividing line between concepts in the data base and look for the side of the line on which the new point falls.

In fact, this is exactly how nearest neighbor works. Consider the points shown in Figure 3. If the points represent columns in the fault dictionary, diagnosis consists of finding the point (called an "exemplar") in the data base that most closely matches the point to be classified (i.e., the test results). Geometrically, the "dividing line" between two exemplars used to determine the nearest neighbor is a line perpendicular to the line connecting the two exemplars. Further this line intersects the connecting line at the midpoint between the two exemplars. This is shown with the same three points in Figure 4. The set of dividing lines between the exemplars (i.e., the dividing hyperplanes in higher dimensions than two) is called a *Voronoi diagram*. Nearest neighbor classification consists of determining in which region the point to be classified falls and assigning the corresponding label.

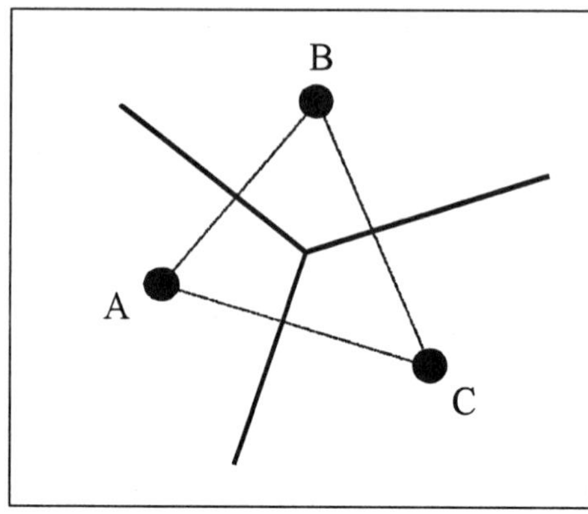

Figure 4. Voronoi diagram for example points.

Assuming the data stored in the dictionary is correct,[4] at least three situations arise in which nearest neighbor classification may result in error. The first case arises when the attributes (e.g., the test results) are irrelevant or insignificant. This situation is illustrated in Figure 5. In this figure, assume only one attribute (corresponding to the x-axis) is significant for classifying the points as either "black" or "white." Next assume an irrelevant attribute has been added to each of the points (corresponding to the y-axis). For all practical purposes, we can assume the second attribute can be modeled as white noise. The vertical line in the figure represents the correct decision boundary between "black" and "white," but the jagged line represents the Voronoi diagram corresponding to the exemplar set. If a point to be classified falls inside any of the regions labeled "error," that point will be mis-classified. In the case base, an irrelevant attribute is analogous to considering an output other than in the system under test. Since we have complete control over which observable outputs to consider, this source of error should not be a problem; however, care should be taken not to include tests that add nothing to diagnostic resolution. Such tests introduce noise since they are "irrelevant."

The second source of error arises when a significant attribute is missing from the exemplar set. For example, suppose the points in Figure 6 are correctly classified along the decision boundary shown. This boundary assumes both attributes (corresponding to both the x- and y-axes) are present. But suppose the fault simulator fails to consider the y-axis. In other words, a particular test vector is not processed by the simulator. This is equivalent to projecting all of the points into the x-axis. In Figure 6, we see the decision boundary then overlaps which leads to another source of error. In the case

---

[4] We will not consider here the case where erroneous data may have been introduced into the exemplar set. Nevertheless, it should be clear that the ability to classify is limited by the quality of the exemplars.

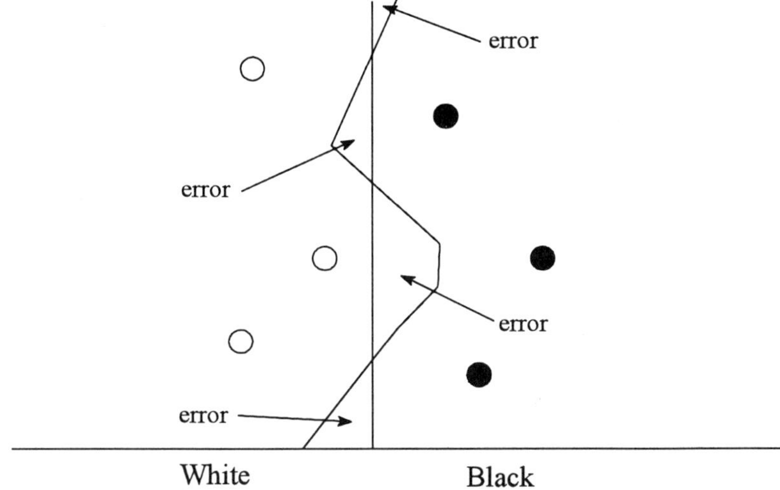

Figure 5 Errors arising from irrelevant attributes.

base, the overlap is equivalent to increasing ambiguity between failure modes, and can arise as a direct result of case base compression when reducing the set of tests.

The third source of error arises from the fact that nearest neighbor can be considered a method for function approximation. In the case where the decision boundaries do not consist of linear segments, the actual boundary can only be approximated by nearest neighbor. In outcome-based testing, all of the stored points exist on the corners of an *n*-dimensional hypercube, so nearest neighbor is able to model the decision boundaries exactly. However, if points are missing from the space of possible points, then it is possible that the decision boundaries will not be modeled correctly. For example, suppose the white point shown in Figure 7 is missing. Then presentation of that point for classification later will result in an error since all points on the lower left side of the decision boundary will be classified as "black." Unfortunately, this is analogous to what happens when compressing the set of fault signatures, except in such a case, the missing point corresponds to a missing failure mode. Thus classification is impossible since a class is completely unrepresented.

5.2   The Appropriateness of Nearest Neighbor

In determining the appropriateness of nearest neighbor for outcome-based diagnosis, we must examine the characteristics of diagnosis and outcome-based case bases. Then we consider the characteristics of the exemplars to be used for diagnosis and look at the impact of these

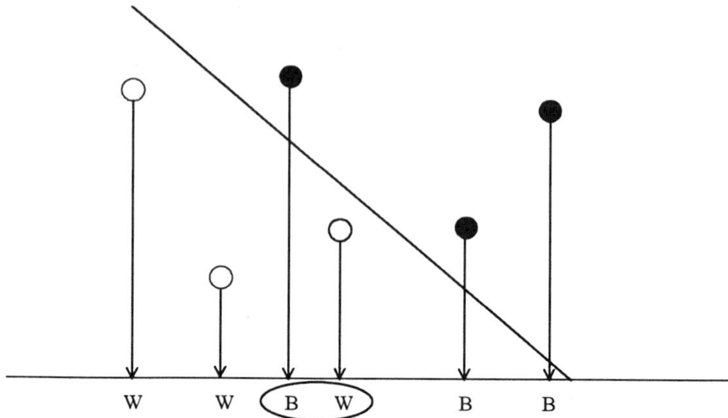

Figure 6. Error arising from missing attributes.

characteristics on the potential for error. Given our discussion on error sources, in this section, we will only consider the case where we may have missing attributes, noisy attributes, or missing exemplars.

When we construct a case base, we assume several characteristics of the diagnostic process. First, we assume we will be able to apply all of the stimuli (i.e., all the tests are available and performable). Further, we assume that the results we obtain from testing reflect the circuit under test (whether failed or not). Finally, we assume that all of the failure modes of interest to us are modeled in the fault represented in the case base.

Nearest neighbor classification is appropriate when the exemplars in the data base are *representative* of possible examples to be classified. The exemplars are intended to model the underlying distributions of the classification space. Cover and Hart (1967) motivated using the nearest neighbor rule with the following:

> If it is assumed that the classified samples (x_i, θ_i) are independently identically distributed according to the distribution of (x, θ), certain heuristic arguments may be made about good decision procedures. For example, it is reasonable to assume that observations which are close together (in some appropriate metric) will have the same classification, or at least will have almost the same posterior probability distributions on their respective classifications.

Unfortunately, the set of exemplars defined by outcome-based testing are not independently identically distributed (i.e., representative of random variables associated with the underlying classes or failure modes) according to the distribution of possible points. This is because most case bases are

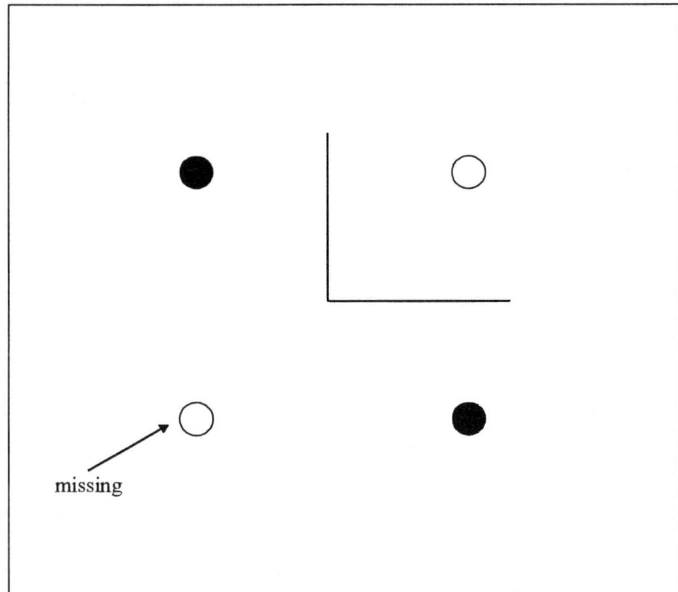

Figure 7. Errors from missing data.

constructed to maximize detection, and the concern is only to provide at least one input vector to detect each failure mode. Further, faults within the system affect more than one test showing some form of dependence.

Other possible reasons outcome-based diagnostics are not well suited for nearest neighbor include the following. For the exemplar set to be effective, it must be representative of the underlying probability distributions. This assumes sufficient sample size for each of the classes (i.e., failure modes) in the case base. Unfortunately, computational complexity precludes generating and using a comprehensive case base. Many tools apply a "detect limit," n, which results in signatures corresponding to some fault being discarded when other signatures detect that fault at least n times. This approach forces under-representation of the classification space which violates the assumption of the points representing the underlying distribution. Also when the underlying distributions of the attributes and classes are not adequately sampled and those distributions are discrete, it is nearly impossible for the nearest neighbor classification procedure to be reliable.

In addition, the exemplars generally are not independent; although, an assumption of independence is necessary for the single stuck-at fault model to apply. Independence breaks down when we consider the effects of one fault on other parts of the system (e.g., it is not uncommon for the presence of a fault to cause another fault). Thus it would appear nearest neighbor is not an appropriate decision rule for outcome-based diagnostics in the general case.

5.3   Nearest Neighbor Diagnosis and the Error Radius

Under very specific conditions, it may be that nearest neighbor will provide good diagnosis when using outcome-based testing. Recent work in network fault management and alarm correlation is analogous to fault-dictionary-based diagnosis (Kliger, Yemini, Yemini, Ohsie, and Stolfo, 1995). In this work, the relationships between test results and failure modes are represented using coding theory where a "codebook" is constructed to include the set of alarms pertinent to diagnosis and the specific failure modes of interest. Diagnosis then consists of applying a nearest neighbor classification rule to the codebook with a set of symptoms to determine the associated failure mode.

In the network fault management problem, the possibility of error in the test results is quite high; therefore, handling erroneous test results in diagnosis is critical. To guarantee correct diagnosis in the presence of noise, Kliger *et al.* (1995) examine the characteristics of the codebook and define precise conditions under which nearest neighbor classification is appropriate.

Central to nearest neighbor being applicable to diagnosis with the case base (or a codebook) is understanding the nature of an error radius. The *error radius* is defined to be the minimal Hamming distance among the code vectors in the codebook. The codebook is constructed in a way similar to constructing a fault dictionary. In particular, each row of the codebook corresponds to a failure mode, and each column corresponds to a symptom. These symptoms are the same as the tests in the fault dictionary. A cell in the codebook receives the value '1' when the associated symptom appears in the presence of the failure mode and '0' otherwise. Thus the fault dictionary can be converted into the codebook by converting the actual outputs of the circuit to '1' or '0' depending on whether the outputs are expected to be in error.

This model can be referred to as a "causality likelihood model." An entry in the matrix can be interpreted as the likelihood that the test will detect the presence of the associated failure mode. In the deterministic case where we assume the model is correct, the likelihood values would either be '0' or '1'. In the more general case, we can associate a value between zero and one corresponding to the probability of the test failing given the presence of the failure mode.

When considering the possibility of error in the fault signature, we wish to match the signature to the codebook recognizing we will not be able to obtain an exact match. In such a case, we would like to know the number of errors in the signature we can tolerate and still find the correct fault. A measure of the tolerance to error is given by the error radius. For example, if the error radius is one, this means two signatures in the codebook differ by exactly one bit. If that bit is in error, we will not be able to distinguish between the two failure modes. From this we can conclude that the higher the error radius, the more resilient the codebook is to noise.

Several important assumptions underlie the coding method and the use of the error radius. The key assumption is that, as with the fault dictionary, the coding method assumes the underlying model is correct and complete. Correctness requires the code vectors to be accurately modeled and that error results in the testing process only. Completeness requires that all relevant failure modes are included and fully represented by the code vectors; errors arising from an unanticipated failure mode are assumed not to occur or not to be relevant to the diagnosis underway.

If in fact the cause of an erroneous fault signature is the test process or the propagation of the signal through the circuit and not the presence of an unanticipated fault in the circuit, then testing within the error radius may be effective. Table 3 shows a codebook corresponding to the circuit in Figure 2 and based on the reduced fault dictionary in Table 2. Given this codebook, we find the error radius is one (if we collapse all ambiguity into a single row in the matrix). Also, of the $2^8 = 256$ possible code vectors, only 13 are represented. Thus this simple circuit will not be able to process noisy fault signatures well at all.

In larger circuits, the number of possible code vectors increases exponentially, yet the number of signatures included in the codebook does not (since it is not generally practical to generate all possible vectors). Unless ambiguity increases, the subset of signatures must result in an error radius less than or equal to the error radius of the complete set. To prove this, note that eliminating a test vector is equivalent to eliminating an attribute in the exemplar set (not a whole exemplar). Comparing exemplars in which the attribute value is different, deletion of the attribute must decrease the Hamming distance, thus potentially decreasing the error radius. Comparing the exemplars in which the attribute value is the same, deleting the attribute has no effect on the Hamming distance, thus not changing the error radius. Thus the potential for error cannot decrease with the smaller attribute set.

6. DIAGNOSIS WITH DIAGNOSTIC INFERENCE MODELS

In Chapter 6, we discuss using a diagnostic inference model (also called an information flow model) to capture relationships between test outcomes and diagnoses (Sheppard and Simpson, 1998). The concept of the diagnostic inference model is closely related to the fault dictionary, except that it abstracts the concepts of a test and a fault to address testing at the "system level" (Sheppard and Simpson, 1991; Simpson and Sheppard, 1994).

Table 3. Codebook for sample circuit.

	t_1	t_2	t_3	t_4	t_5	t_6	t_7	t_8
a_0	0	1	1	1	0	0	0	1
a_1	1	0	0	0	1	1	1	0
b_1	0	0	1	0	1	0	0	0
c_1	0	1	0	0	0	0	1	0
d_1	1	0	0	0	0	0	0	0
f_0	1	0	0	1	0	0	0	0
f_1	0	1	1	0	1	1	1	1
g_1	0	1	1	0	0	0	0	1
i_0	0	0	0	0	1	1	1	0
i_1	1	0	0	0	0	0	0	0
j_1	0	1	1	0	0	0	0	1
k_0	0	0	0	1	0	0	0	0
k_1	1	1	1	0	0	0	0	1
m_0	0	0	0	1	1	1	1	0
nf	0	0	0	0	0	0	0	0

Inherently, diagnostic inference models are difficult to construct. This difficulty arises from the fact that DIMs require the definition of a comprehensive set of tests, diagnoses (usually based on known failure modes) and relationships between the two. Further, special inference types (e.g., asymmetric inference or cross-linked inference) may be required to capture accurately the inferences derivable from the test results. Unfortunately, such difficulty increases the potential for inefficient or even incorrect diagnosis. Further, as systems increase in complexity, the likelihood of erroneous models increases. Two questions naturally follow from this problem: 1) How does one develop models that minimize the chance of an error? 2) If errors occur, how does one identify and correct them?

Results from machine learning research suggest potential answers to both questions. The most common approach is to apply simulation or fault insertion to generate examples that capture failed behavior and determine test-to-fault relationships. This is probably the most reliable approach to learning or constructing a model; however, it tends to take an inordinate amount of time before useful models are constructed. Related to this approach is using historical data (when available) to construct the model. The next section discusses an approach to do just that.

7. CONSTRUCTING A DIAGNOSTIC INFERENCE MODEL FROM CASE DATA

The primary goal of this section is to discuss methods for constructing diagnostic inference models from case data used in a CBR system. A straightforward approach to doing this, but one that is not likely to generalize well is to treat each case as a unique class (i.e., diagnosis). Then the signature for that class would correspond to the feature values associated with the case. For multi-value features, the binary equivalents in the DIM would be a pairing of the feature label with the feature value. For example, suppose we have a case base with the following three cases.

Case	Test 1	Test 2	Test 3	Test 4
c1	Pass	Fail	Unknown	Unknown
c1	Unknown	Fail	Fail-Lo	Unknown
c2	Fail	Unknown	Fail-Hi	Pass

Note that two of the cases yield the same diagnosis, and one of the tests has three possible outcomes (assuming all tests can pass). Since the DIM assumes a test dependency indicates detection of a fault, an indication of a test passing just means that the test does not depend on the associated fault. Unknown outcomes were not evaluated but may have been inferred.

If we treat each case as a separate class, or diagnosis, then this set of cases would result in the following model:

Case	Test 1	Test 2	Test 3-Lo	Test 3-Hi	Test 4
c1-a		X			
c1-b		X	X		
c2	X			X	

The problem with this approach is that diagnosis would occur (with high confidence) only when a signature is matched exactly (i.e., when one of the cases is experienced again). But this means there is no facility for generalization. This is not a desirable feature, so we would like to provide a method for "combining" cases with the same diagnosis to yield greater generality.

Consider the same three cases. We still need to map multiple-outcome tests into a set of several binary tests, so the columns would remain the same. We only have one case for c2; therefore, we would retain this case in the model as in the previous example. But we should be able to combine the two cases for c1 into a single signature in the model. To do this, we need to ensure that the two cases do not conflict. If they do conflict, then we are back to creating separate signatures.

For this example, the cases do not conflict. This is clear because all of the tests whose values are known for all of the cases agree. If they disagreed, then there would be a conflict. Since there is no conflict, the simplest process for building the signature is to take the union of the signatures from the first model. Doing this yields,

Case	Test 1	Test 2	Test 3-Lo	Test 3-Hi	Test 4
c1		X	X		
c2	X			X	

The problems arise when there are conflicts in the test results. Previously, we provided an intuitive "definition" of conflict as occurring when two cases with the same diagnosis have attributes with different values. Since the case base is not supposed to hold any incorrect cases (i.e., cases where the either the diagnosis or the attribute values were in error), we should be safe in assuming that these conflicts arise only when the attribute can legally take on *either* value for that diagnosis.

Consider the following example. In some sense, it is the simplest example in that it reveals a direct conflict for the same diagnosis. This conflict is on Test 1.

Case	Test 1	Test 2	Test 3	Test 4
c1	Pass	Fail	Unknown	Unknown
c1	Fail	Unknown	Pass	Unknown

For the other three tests, we can simply take the union as we did before. For Test 1, since it can either pass or fail, we need to either treat these as two different conclusions (as we did in the naive example above), or we need to consider what could lead to the conflict. Assuming the tests are correct and reliable (which is a standard assumption when constructing DIMs), we can assume Test 1 is actually *asymmetric*. In particular, we note that Test 1 detected c1 in the second case, but when Test 1 passed, we were not able to rule out c1. This would yield a single signature of

Case	Test 1 (–)	Test 1 (+)	Test 2	Test 3	Test 4
c1	X		X		

A brute-force approach for handling this type of conflict when constructing the DIM is to declare all tests in the model to be *fully asymmetric*. This means that the positive and negative inference lists may be different. Next, group the cases such that all cases with the same diagnosis are in the same group. For a particular group, define a signature in the DIM. Consider all of the failed tests first. If a case exists in the group with a failed test, enter the diagnosis as a dependency of that test on both the positive and negative inference side. Next consider all of the passed tests. If a case exists in the group with a passed test where a dependency has been entered, remove that dependency on the positive inference side. Do not worry about tests whose values are unknown—they will be treated as not depending on the fault until evidence to the contrary is encountered. Finally, after the model is constructed, the inference lists can be compared. If any test has the same list for both the positive and negative inference sides, that test can be converted back into a symmetric test.

Now consider the case where we have a multiple-outcome test (e.g., Test 3 with possible values of pass, fail-hi, and fail-lo). This may yield a "conflict" such as the following:

Case	Test 1	Test 2	Test 3	Test 4
c1	Pass	Unknown	Fail-Lo	Unknown
c1	Unknown	Fail	Fail-Hi	Unknown

As before, this situation indicates the test can take on two legal values in the presence of this fault. The difference, however, is that the same "test" succeeds in detecting the fault. If we construct the model as before, we will still yield non-conflicting signatures. Thus,

Case	Test 1	Test 2	Test 3-Lo	Test 3-Hi	Test 4
c1-a		X	X		
c1-b		X		X	

would be an acceptable model. Note that the test failure for Test 2 is entered into both signatures since it is not a point of conflict. In this case, it is also desirable to create a *linked outcome* between Test 3-Lo and Test 3-Hi such that if either fails, the other must pass. This is because it is impossible for Test 3 to have the value of Fail-Lo and Fail-Hi at the same time. But this leads to a pass/fail conflict similar to above.

To combine these two signatures, we would have to treat each "side" of Test 3 as asymmetric as we did before. Thus, Test 3-Lo and Test 3-Hi would both be fully asymmetric tests, but c1 would be absent from the

dependency list of the positive inference sides of both tests. Then we would
have a single signature.

linked

Case	Test 1	Test 2	Test 3-Lo (–)	Test 3-Lo (+)	Test 3-Hi (–)	Test 3-Lo (+)	Test 4
c1		X	X		X		

For cases where the structure of the case is more complex than a
simple "feature vector," the structure would need to be flattened into the
feature-vector form to facilitate mapping into the DIM. This should be
straightforward since most case representations that are not flat use the
structure to provide means of speeding up search and comparison. Leaf
values in the structure (possibly with some propagation downward through
the case structure) should be sufficient for constructing the feature vectors.

It should be apparent that the completeness of the model that results
from mapping the case data depends heavily on the richness of the cases
stored in the case base. Areas of the case base that do not adequately
represent the corresponding set of faults may lead to situations in which the
model-based approach leads to inaccurate or imprecise results. This is not
unexpected. In fact, similar problems are likely to result when applying the
CBR system should test results lead to that part of the case base. Recalling
the observation by Cover and Hart, we see why it is important to have a
representative sample of case data for performing diagnosis with CBR or for
inducing an DIM from the case data.

8. REASONING UNDER UNCERTAINTY WITH DIMS

By now it should be apparent that great care should be used when
diagnosing faults with an outcome-based approach. Because of the inherent
difficulties in processing erroneous test data in these approaches, we
developed two alternative approaches to processing this data which are
presented in Chapter 6 (Sheppard and Simpson, 1998). The alternatives
consider test results as evidence for or against the presence of a fault in the
system. Test results are processed sequentially, and the evidence supporting
or denying the presence of a failure mode is attributed to the set of failure
modes in the system. These approaches are based on the Dempster-Shafer
method for reasoning under uncertainty (Dempster, 1968; Shafer, 1976) and
certainty factors as incorporated in the *MYCIN* system (Shortliffe, 1976).

In related research, Denœux (1995) attempted to overcome some of the drawbacks of the nearest neighbor classification rule and proposed a modification to k-nearest neighbor using Dempster-Shafer. Denœux notes that "the main drawback of the voting k-NN rule is that it implicitly assumes the k nearest neighbors of a data point x to be contained in a region of relatively small volume, so that sufficiently good resolution in the estimates of the different conditional densities can be obtained." He goes on to point out that in practice, "the distance between x and one of its closest neighbors is not always negligible, and can even become very large outside the regions of high density." Since an outcome-based case base defines points on the corners of an n-dimensional hypercube, it is reasonable to assume this "pathological" condition holds.

The proposed solution is to gather "evidence" from the neighbors and use this evidence to classify the point x. The basic method Denœux uses is to identify the k nearest neighbors and combine the evidence provided by these neighbors for each of the classes under consideration. Specifically, he compute the "basic probability assignments" (BPA) as $\mu^{s,(i,j)} = \mu^{s,i} \oplus \mu^{s,j}$ where s denotes the sample to be classified, and x^i and x^j are two points which are in the set of nearest neighbors Φ^s that belong to the same class C_q. The BPAs are computed using sets of nearest neighbors of each class, Φ_q^s, and the associated evidence is combined for each class. Using this formulation, the rankings provided by both support and plausibility are consistent, so the class assigned to x^s is simply the class with the maximum support (or plausibility).

While providing improvement over the standard k-nearest neighbor rule, all of the problems described previously still apply. While this approach tends to reduce the reliance of diagnosis on homogeneous regions of points in the instance base, it still assumes neighboring regions provide information about the point to be classified. This is unlikely to hold in outcome-based diagnostics, and we believe the evidence to be provided for classification (i.e., diagnosis) comes from *test* information rather than neighboring *diagnoses*. This approach is fundamental to the diagnostic inference modeling approach.

9. DIAGNOSIS OF A SIMPLE CIRCUIT

In this section, we will walk through an example diagnosis with the circuit in Figure 2 and illustrate how an inference method such as the Dempster-Shafer approach can still provide a reasonable answer in the presence of uncertainty, even when nearest neighbor classification does not. First we will consider the performance of nearest neighbor on the sample circuit, and then we will apply Dempster-Shafer inference with the diagnostic inference model on some example diagnoses. For this example, we must

Table 4. Accuracy using nearest neighbor on a fault dictionary.

Bit Errors	1	2	3	4	5	6	7	8
Hamming Dist.	82	110	42	9	0	0	0	0
correct diag.	79%	30%	6%	1%	0%	0%	0%	0%
incorrect diag.	21%	70%	94%	99%	100%	100%	100%	100%
Overlap	76	91	57	19	0	0	0	0
correct diag.	73%	25%	8%	2%	0%	0%	0%	0%
incorrect diag.	27%	75%	92%	98%	100%	100%	100%	100%
Total Cases	104	364	728	910	728	364	104	13

assume we have received an erroneous fault signature. We will need to define confidence values for the test results and process the test results through the model.[5]

9.1   Diagnosis with Nearest Neighbor

First, we will consider the ability of nearest neighbor to process erroneous fault signatures in a fault dictionary. For the following demonstration, we considered only the sample circuit and used both Hamming distance and the overlap metric. The matching procedure was limited to 1-NN, and we expect worse results for k-NN with $k > 1$ since faults are only represented by one signature each. For each fault, we considered all possible fault signatures that can be generated with one through eight bits in error. We then compared the results of using 1-NN with the expected fault and recorded the number of correct diagnoses. The results of these experiments are given in Table 4.

From this table, we see some characteristics of introducing error into the fault signature to be matched. First, we see that the higher the number of bits in error, the lower the accuracy in matching, down to a limit of 0% accuracy. Second, the performance of Hamming distance compared to the overlap metric is very close. In fact, we conjecture that the differences are not statistically significant; although, we do not have sufficient data to perform a significance test. Third, the lowest error rate (i.e., one bit error) yielded very poor performance on this circuit (between 21% and 27% error). This should not be a surprise given the previous discussion on the appropriateness of nearest neighbor, but it may be disconcerting to those who apply nearest neighbor in their diagnostic systems.

---

[5] Our algorithm for applying Dempster-Shafer includes calculation for support of an "unanticipated result." This special conclusion is included to address issues related to conflicting test results. Fortunately, the presence of conflict does not affect the relative positions of the conclusions in the table since support for the unanticipated result only modifies the normalizer for other support values.

Given the poor performance on nearest neighbor, we will take two cases in which nearest neighbor fails to find the correct failure mode and process those cases with the diagnostic inference model and the Dempster-Shafer methodology. We will then reconstruct Table 4 using Dempster-Shafer. The first case will use a fault signature with one bit in error, and the second case will use a signature with two bits in error. Given the rapid degradation after two bits (since more than 25% of the bits are now wrong), we will assume Dempster-Shafer will be affected similarly with the high error percentages.

In selecting our two test cases, we want nearest neighbor to fail to find the correct fault. Since we know the number of incorrect bits, we will identify two failure modes in the fault dictionary whose Hamming distance corresponds to twice the number of incorrect bits minus one and ensure the proper number of differentiating bits are in error. This will result in selecting the wrong failure mode, but with a signature that is not in the fault dictionary.

9.2   One-Bit Error with Dempster-Shafer

For the first case, we note that the Hamming distance between c_1 and *nf* is two. In other words, two tests detect the presence of c_1, and if both of those tests are in error, no fault will be detected. This is a common occurrence in testing, and effective diagnostics as well as effective testing are necessary to combat this problem. For our example, we will assume one of the two tests capable of detecting c_1 fails to make the detection (i.e., the test passes). Without loss of generality, we will select t_2 to pass when it should fail. We will also assume sufficient test data has been gathered to identify t_2 as a problem test, and we will reduce its confidence to 0.75. All other test confidences will be set to 0.99

The results of processing all eight test vectors through the Dempster-Shafer calculations with the diagnostic inference model are given in Table 5. The normalized values for evidential probability, though quite low, show the leading candidates for diagnosis are c_1 and *nf*. This result is consistent with what we would expect given the nature of the test data.

Another interesting result when considering the test data applied to the diagnostic inference model came when we omitted t_2 from the calculations. Although both c_1 and *nf* were still the leading candidates, the differences in probabilities between c_1 and *nf* were such that c_1 could be declared with greater confidence to be the fault. In fact, this makes sense given the new test result from t_2 contradicts c_1 as the hypothesis.

Table 5. Dempster-Shafer calculations with one bit error.

Failure Mode	Support	Plausibility	Probability
a_0	0.043	0.411	0.0505
a_1	0.079	0.629	0.0767
b_1	0.067	0.629	0.0768
c_1	**0.111**	**0.906**	**0.1057**
f_0	0.070	0.629	0.0760
f_1	0.052	0.411	0.0514
i_0	0.093	0.753	0.0902
i_1	0.084	0.753	0.0895
j_1	0.056	0.535	0.0652
k_0	0.084	0.753	0.0895
k_1	0.043	0.411	0.0505
m_0	0.079	0.629	0.0767
nf	**0.098**	**0.876**	**0.1024**

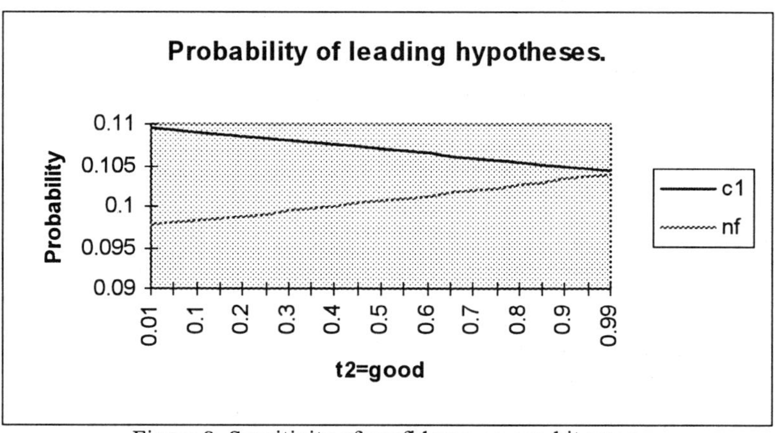

Figure 8. Sensitivity of confidence on one-bit error.

Given this result, we decided to perform a sensitivity analysis on the erroneous test to determine at what point the incorrect diagnosis would be returned. For this analysis, we varied the confidence in the test outcome from 0.01 to 0.99 and observed the relative difference in probability of the correct answer (i.e., c_1) and the nearest neighbor (i.e., *No Fault*). We expected the choice from Dempster-Shafer to flip at some point and return the incorrect answer. As we can see from Figure 8, however, this did not happen. In fact, we found that Dempster-Shafer returned the correct example regardless of the confidence value; however, as confidence increased, the difference between c_1 and *No Fault* converged to the point they were no longer significantly different. In fact, the hypothesis generation routine returned both faults as possible answers.

9.3　Two-Bit Error with Dempster-Shafer

For the second test case, we identified two failure modes whose Hamming distance was three. This indicated only three test results distinguished the two conclusions. Given the application of nearest neighbor with two tests results in error when the two tests were among the three distinguishing the conclusions, we would expect the wrong conclusion to be identified.

Assume the failure mode present in the system is a_1. Tests t_5, t_6, and t_7 differentiate this failure mode from failure mode i_1. Without loss of generality, suppose both t_6 and t_7 are in error. Again, we assume we have sufficient test data to warrant assigning confidence values of 0.75 to these two tests and 0.99 for all other tests. The results of processing all eight test vectors through the Dempster-Shafer calculations with the diagnostic inference model are given in Table 6. The normalized values for evidential probability, this time, show the leading candidates for diagnosis are a_1 and i_0. Again, this result is consistent with what we would expect given the nature of the test data since the Hamming distance between a_1 and i_0 is only 1.0. The erroneous conclusion drawn by nearest neighbor of i_1, though third, never appears as a member of the hypothesis set.

As before, we examined the results of the Dempster-Shafer calculations without the "conflicting" test results from t_6 and t_7. No conflict was evident without these test results. Further, we once again found that the hypothesis without these two tests was a_1 by itself. It was only in the presence of the conflicting test results that i_1 was added to the hypothesis, but a_1 remained the preferred failure mode. Also, when only t_6 was included, though conflict was now present in the test results, the conflict was not sufficient to add another failure mode to the hypothesis; a_1 remained the sole failure mode in the hypothesis set.

We also conducted a sensitivity analysis on the confidence values of the two erroneous tests. This time, we varied the confidence values from 0.01 to 0.99 in a factorial study. The results of this analysis are given in Figure 9 and represent the difference in probability between a_1 and i_0. As before the correct fault is always identified as the first choice using Dempster-Shafer, but this time, as confidence increases, the difference does not become as small.

Table 6. Dempster-Shafer calculations with two bit errors.

Failure Mode	Support	Plausibility	Probability
a_0	0.035	0.258	0.0392
a_1	**0.131**	**0.813**	**0.1180**
b_1	0.066	0.535	0.0790
c_1	0.029	0.318	0.0465
f_0	0.073	0.535	0.0798
f_1	0.093	0.535	0.0820
i_0	**0.100**	**0.689**	**0.1012**
i_1	0.087	0.659	0.0966
j_1	0.049	0.381	0.0574
k_0	0.043	0.411	0.0606
k_1	0.080	0.505	0.0768
m_0	0.087	0.565	0.0850
nf	0.056	0.535	0.0779

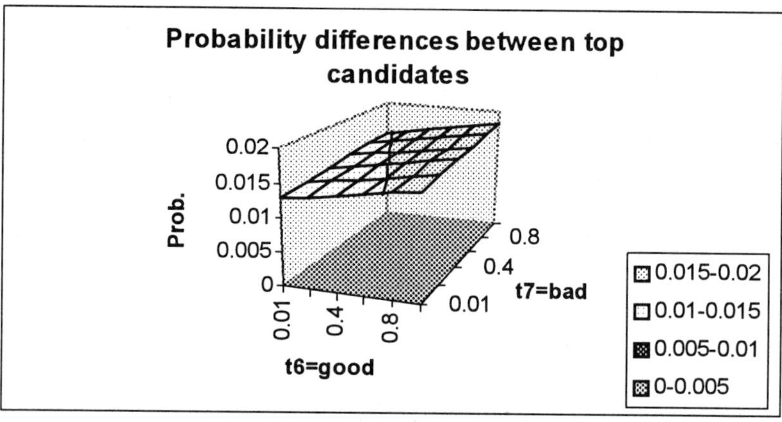

Figure 9. Sensitivity of confidence on two-bit error.

9.4   Dempster-Shafer and Nearest Neighbor Compared

To further compare the differences between the Dempster-Shafer approach and nearest-neighbor classification, we computed the accuracy for all bit-error combinations using Dempster-Shafer as we did for nearest neighbor. These results are shown in Table 7. In interpreting this table and Table 4, we can consider the bit errors as corresponding to some amount of lost information. For example, in the two-bit error case, we assume 25%

Table 7. Accuracy using Dempster-Shafer on a fault dictionary.

Bit Errors	1	2	3	4	5	6	7	8
Correct = 1st	89	174	150	62	0	0	0	0
correct diag.	86%	48%	21%	7%	0%	0%	0%	0%
incorrect diag.	14%	52%	79%	93%	100%	100%	100%	100%
Correct = $1^{st}/2^{nd}$	104	331	336	66	0	0	0	0
correct diag.	100%	91%	46%	7%	0%	0%	0%	0%
incorrect diag.	0%	9%	54%	93%	100%	100%	100%	100%
Total Cases	104	364	728	910	728	364	104	13

information loss. From this we can see that even one bit error is significant in that it corresponds to 12.5% information loss.

Consider the rows labeled "Correct = 1st." These rows correspond to the analysis when we consider the conclusion assigned the highest probability of being correct. This is analogous to the nearest-neighbor case in which we select the fault whose signature is closest to the test signature as the most likely diagnosis. Comparing these rows with Table 4, we find that Dempster-Shafer strongly outperforms both Hamming distance- and Overlap metric-based nearest neighbor. In fact, we see that with 37.5% information loss, nearest neighbor performs randomly (i.e., if we randomly select from the 13 possible failure modes meaning any failure mode might be selected with probability 7.7%, we will be correct approximately the same number of times as nearest neighbor with three bits in error). On the other hand, Dempster-Shafer does not reduce to "random" performance until we have 50% information loss. When information loss exceeds 50%, both techniques fail to find the correct diagnosis, and this is not unexpected.

An interesting result with Dempster-Shafer involves examining the number of times the correct answer is either the first or second most likely conclusion identified (shown in the rows labeled "Correct = $1^{st}/2^{nd}$"). Here we find the correct fault a very high percentage of the time, indicating an alternative answer in the event repair based on the first choice is ineffective. In fact, in all cases where the answer was ranked either first or second, Dempster-Shafer still considered it to be a member of the hypothesis set.

A closer examination of the results using Dempster-Shafer yield some interesting observations. If we limit our consideration to the one-bit error case, we find that Dempster-Shafer returns the correct diagnosis as either the first or second choice in all cases. Examining the tests that are in error, we find an interesting pattern. In all cases where the correct diagnosis is second, either t_1 or t_4 is in error, and in five out of the eight cases, both of these tests in error result in the correct diagnosis being listed second. For the other three cases, one of the tests result in the "wrong" answer, but the other does not.

In all cases, we can explain the cause of the error by examining the Hamming distance between the reported most likely diagnosis and the correct diagnosis. In these cases, the Hamming distance is one, and the bit that is different is the bit whose error leads to the wrong diagnosis. This supports the conclusion of the significance of the error radius and seems to indicate that, as long as the number of bits in error is less than the error radius of the signatures in the fault dictionary, Dempster-Shafer will yield the correct result.

We can extend this result by examining the relative Hamming distances between all of the faults in the fault dictionary. Those faults with a large amount of similarity will be expected to lead to incorrect diagnosis in the presence of higher bit errors in the signature. In fact, this is exactly what we observed. With both the two- and three-bit error cases, we found faults i_0, i_1, j_1, and k_0 had difficulty, but the number of other signatures similar to these four signatures (i.e., with Hamming distance less than or equal to three) was high. This further confirms the significance of the information provided by the tests and its ability to distinguish faults in contrast to focusing on the conclusion landscape to determine proper diagnoses.

10.   CONCLUSION

In this chapter, we described an approach to generating diagnostic inference models from case data stored in a CBR system. We also provided an approach to reasoning under uncertainty using the resultant model. The rationale for developing this approach was two-fold. First, we wanted to improve the diagnostic process by providing a more compact representation of the diagnostic knowledge. Second, we wanted to provide a mechanism whereby understanding the diagnostic knowledge was possible. CBR systems are typically very slow where DIM-based systems are fast. Further, it is difficult to develop any understanding of the structure and meaning of knowledge implicit in the cases of a CBR system where model-based systems emphasize understanding this structure. Third, given degraded accuracy resulting from case-based diagnosis applied to outcome-based testing, we wanted to provide a means for improving accuracy using the derived models.

The primary disadvantage to any model-based system (including DIM-based systems) is the difficulty in developing the model. While, conceptually, the elements required for DIMs are easy to understand, the process of collecting and synthesizing the data into the DIM is highly labor-intensive and error-prone. It is difficult to dispute the accuracy of case data since it corresponds to actual diagnostic experience (unless erroneous case information is stored). Therefore, the advantage to processing the case data to generate an DIM is that is provides a solid foundation of experience from which to derive the models. The resulting system then enjoys the advantages of both case-based systems and model-based systems while simultaneously minimizing the effects of the disadvantages.

11. ACKNOWLEDGMENTS

The work presented in this chapter combines research in diagnostic and machine learning. Consequently, it reflects the advice and counsel of several people. Thanks go to Jerry Graham, Don Gartner, Steven Salzberg, David Aha, Harry Dill, and Warren Debaney. Without their insight and advice, this work would not have been completed.

12. REFERENCES

Aamodt, A. and E. Plaza. 1994. "Case-Based Reasoning: Foundational Issues, Methodological Variations, and System Approaches," *AI Communications*, Vol. 7, No. 1, pp. 39–59.

Abramovici, M., M. A. Breuer, and A. D. Friedman. 1990. *Digital Systems Testing and Testable Design*, New York: Computer Science Press.

Cover, T. M., and P. E. Hart. 1967. "Nearest Neighbor Classification," *IEEE Trans. on Information Theory*, Vol. IT-13, pp. 21–27.

Dasarathy, B. V. (ed.). 1991. *Nearest-Neighbor Norms: NN Pattern Classification Techniques*, Los Alamitos, California: IEEE Computer Society Press.

Debaney, W. H. and C. R. Unkle. 1995. "Using Dependency Analysis to Predict Fault Dictionary Effectiveness," *AUTOTESTCON '95 Conference Record*, New York: IEEE Press.

Denœux, T. 1995. "A *k*-Nearest Neighbor Classification Rule Based on Dempster-Shafer Theory," *IEEE Transactions on Systems, Man, and Cybernetics*, Vol. SMC-25, No. 5, pp. 804–813.

Devijver, P. A. and J. Kittler. 1982. *Pattern Recognition: A Statistical Approach*, Englewood Cliffs, New Jersey: Prentice-Hall.

Dempster, A. P.. 1968. "A Generalization of Bayesian Inference," *Journal of the Royal Statistical Society*, Series B, pp. 205–247.

Grant, F. 1986. "Noninvasive Diagnostic Technique Attacks MIL-SPEC Problems," *Electronics Test*, Miller-Freeman Publications.

Kliger, S., S. Yemini, Y. Yemini, D. Ohsie, and S. Stolfo. 1995. "A Coding Approach to Event Correlation," *Fourth International Symposium on Integrated Network Management*, Santa Barbara, California.

Mingers, J. 1989. "An Empirical Comparison of Pruning Methods for Decision Tree Induction," *Machine Learning* Vol. 4, No. 2, pp. 227–243.

Quinlan, R. 1986. "Induction of Decision Trees," *Machine Learning*, Vol. 1, pp. 81–106.

Richman, J. and K. R. Bowden. 1985. "The Modern Fault Dictionary," *Proceedings of the International Test Conference*," Los Alamitos, California: IEEE Computer Society Press.

Ryan, P. 1994. *Compressed and Dynamic Fault Dictionaries for Fault Isolation*, Ph.D. Thesis, University of Illinois at Urbana-Champaign.

Shafer, G. 1976. *A Mathematical Theory of Evidence*, Princeton, New Jersey: Princeton University Press.

Sheppard, J. W. 1996. "Maintaining Diagnostic Truth with Information Flow Models," *AUTOTESTCON '96 Conference Record*, New York: IEEE Press.

Sheppard, J. W. and W. R. Simpson. 1991. "A Mathematical Model for Integrated Diagnostics," *IEEE Design and Test of Computers*, Vol. 8, No. 4, Los Alamitos, California: IEEE Computer Society Press, pp. 25–38.

Sheppard, J. W. and W. R. Simpson. 1998. "Managing Conflict in System Diagnosis," *System Test and Diagnosis: Research Perspectives and Case Studies*, eds. J. W. Sheppard and W. R. Simpson, Norwell, Massachusetts: Kluwer Academic Publishers, 1998.

Shortliffe, E. H. 1976. *Computer Based Medical Consultations: MYCIN*, Elseview, New York.

Simpson, W. R. and J. W. Sheppard. 1994. *System Test and Diagnosis*, Norwell, Massachusetts: Kluwer Academic Publishers.

Stanfill, C. and D. Waltz. 1986. "Toward Memory-Based Reasoning," *Communications of the ACM*, Vol. 29, No. 12, pp. 1213–1228.

Tomek, I. 1976. "An Experiment with the Edited Nearest Neighbor Rule." *IEEE Transactions on Systems, Man, and Cybernetics*, Vol. SMC-6, No. 6, pp. 448–452.

Tulloss, R. E. 1978. "Size Optimization in Fault Dictionaries," *Proceedings of the Semiconductor Test Conference*, Los Alamitos, California: IEEE Computer Society Press, pp. 264–265.

Tulloss, R. E. 1980. :Fault Dictionary Compression: Recognizing when a Fault may be Unambiguously Represented by a Single Failure Detection," *Proceedings of the International Test Conference*, Los Alamitos, California: IEEE Computer Society Press, pp. 368–370.

Wilson., D 1972. "Asymptotic Properties of Nearest Neighbor Rules Using Edited Data," *IEEE Transactions on Systems, Man, and Cybernetics*, Vol. SMC-2, No. 3, pp. 408–421.

Chapter 6

Accurate Diagnosis through Conflict Management

John W. Sheppard
ARINC Incorporated

William R. Simpson
Institute for Defense Analyses

Keywords: Diagnostic inference models, diagnostics, system test, certainty factors, Dempster-Shafer, conflict management.

Abstract: Advanced diagnostic techniques have become essential for supporting reliable testing of modern systems due to the rapid growth in complexity of these systems. Errors in the test process become more likely with increased complexity and need to be treated carefully to maximize accuracy in diagnosis. In this chapter, we discuss several approaches to analyzing conflicting test results. We have concentrated on one approach based on model-based diagnosis—the diagnostic inference model—to identify conflict and improve diagnostics.

1.   INTRODUCTION

The complexity of modern systems has led to new demands on system diagnostics. As systems grow in complexity, the need for reliable testing and diagnosis grows accordingly. The design of complex systems has been facilitated by advanced computer-aided design/computer-aided engineering (CAD/CAE) tools. Unfortunately, test engineering tools have not kept pace with design tools, and test engineers are having difficulty developing reliable procedures to satisfy the test requirements of modern systems.

Testing of complex systems is rarely perfect. In software, it is almost impossible to track system state or all of the ways values might be affected. Hardware systems are frequently subject to noise and other random events making interpretation of test results difficult, thus lowering confidence in what the tests indicate. Even with digital testing, which eliminates some noise problems, developers must still contend with the effects of state. Finally, modern systems depend heavily on both hardware and software, and the interactions between hardware and software further compound the problem of managing test errors.

When testing a complex system, what is the proper response to unexpected and conflicting test results? Should the results be scrapped and the tests rerun? Should the system be replaced or redesigned? Should the test procedures be redeveloped? Determining the best answers to these questions is not easy. In fact, each of these options might be more drastic than necessary for handling conflict in a meaningful way.

When test results conflict, the potential source of the conflict must be analyzed to determine the most likely conclusion. To date, test systems have done little more than identify when a conflict exists. Since the early 1970s, artificial intelligence researchers have attempted to "handle" uncertain and conflicting test information. But handling the uncertainty and the conflict has been limited to assigning probabilities or confidence values to the conclusions to provide a ranked list of alternatives for taking appropriate action (Pearl, 1988; Peng and Reggia, 1990). When test results are uncertain but consistent, this is about the best we can do.

In this article, we discuss two approaches to system diagnosis that apply several tests and interpret the results based on an underlying model of the system being tested. These tests are used to determine if the system is functioning properly and, if not, to explain the faulty system performance. When test information conflicts, ignoring or improperly handling this conflict will degrade diagnostic accuracy. By examining potential sources of conflict and the way conflict might become manifest in a reasoning system, we developed an approach to extend diagnosis to handle the conflict and draw more reliable conclusions.

2. SYSTEM DIAGNOSIS

Frequently, test engineers define a system-level diagnostic process that is independent of the design and manufacturing process. The first step, for example, is to develop built-in test (BIT) or built-in self test (BIST) for initial detection and localization of faults. These tests, which are embedded in the system itself, when used with other tests, may localize faults to a level sufficient to take action. Subsequent steps apply a battery of automatic and manual tests to the system (or subsystem). Eventually, these tests might identify the subunit suspected of containing the fault. The subunit is then tested to find the faulty unit. Once a unit or subunit is separated from the system, maintainers frequently use specialized equipment (usually from the unit manufacturer) to test it.

Despite improvements in BIT, BIST, and automatic testing, manufacturers typically have not provided maintainers with comprehensive diagnostic procedures. Instead, they rely on part screening and special test approaches, which are inadequate in that they emphasize ensuring the system functions properly rather than isolating faults when the system does not function properly

This approach to system testing is an artifact of a manufacturing process that tests only pieces of systems. This approach is clearly insufficient to explain anomalous behavior at the system level, since it fails to account for the complex interactions among system components. At this level, we are left with a few euphemisms to heuristic approaches, such as "tickle testing" (when we snug all fittings and clean all contacts), or "shotgun maintenance" (when we guess where the fault resides and take action until the system anomalies disappear).

In developing an alternative, we focused on ideas developed in *integrated diagnostics* programs. Integrated diagnostics programs emphasize the application of structured approaches to system testing and diagnosis. They have three objectives:

- Maximum reuse of design and test data, information, knowledge, and software.
- Integration of support equipment and manual testing, to provide complete coverage of diagnostic requirements.
- Integration of available diagnostic information, to minimize required resources and optimize performance.

Our research focuses on applying a uniform method for representing diagnostic information: One model type represents the system at all levels of detail. Using this model, test engineers can determine BIT requirements, define test programs for automatic test equipment, and guide the manual troubleshooting process.

The model we use captures test information: It models the information provided by a set of tests defined for the system with respect to a set of desired conclusions. During troubleshooting, the information gathered from performing the series of tests is combined to make a diagnosis. Defining the relationships between tests and conclusions results in the *diagnostic inference model*. The models are hierarchical, in that a conclusion in one model can be used to invoke a lower-level model. The rules for handling each model and submodel are the same regardless of position in the hierarchy.

We begin by developing a set of diagnostic inference models for the system to be tested. We develop models for on-board diagnosis (thus determining the requirements for BIT) and for each subsequent level of testing. The conclusions drawn at one level determine the appropriate model to use at the next level.

Once developed, we analyze the models to evaluate the system's testability and perform design trade-offs to improve testability. Thus the modeling process begins in the early stages of system development. As the system progresses through the life cycle, the models are revised to reflect changes and refinements in the design. For diagnosis, the models define available tests and inferences that can be drawn by obtaining test outcomes. Hence, the same models used to evaluate testability of the system can be used for troubleshooting.

3.   THE DIAGNOSTIC INFERENCE MODEL

To address several problems associated with performing system diagnosis and analyzing system testability, we introduced the concept of an information flow model (Simpson and Sheppard, 1994). More recently, this kind of model has become known as the diagnostic inference model. This model-based approach to system test and diagnosis incorporates techniques from information fusion and model-based reasoning to guide analysis. The model represents the problem to be solved as a set of diagnostic inferences. Tests provide information, and diagnostic inference combines information from multiple tests using information fusion and statistical inference. The structure of the diagnostic inference model then facilitates our ability to compute testability measures and derive diagnostic strategies.

A diagnostic inference model has two primitive elements: *tests* and *fault-isolation conclusions*. Tests include any source of information that can be used to determine the health state of a system. Fault isolation conclusions include failures of functionality, specific non-hardware failures (such as bus timing), specific multiple failures, and the absence of a failure indication (*No Fault*). The information obtained may be a consequence of the system operation or a response to a test stimulus. Thus, we include observable symptoms of failure processes in the diagnostic inference model as tests. Including these symptoms allows us to analyze situations that involve information sources other than formally defined tests. Of course, the purpose of the model is to combine information obtained from these information sources (tests) to derive conclusions about the system being diagnosed.

When developing a fault isolation strategy, the type, amount, and quality of test information should be considered. For our purposes, we initially assume equal quality among test results in the sense that the *good* or *bad* indication of a test actually reflects the state of the unit under test. During actual diagnosis, we relax this assumption to allow a confidence value to be associated with a test result. If all test inferences in a system are known, the information content of each test can be calculated. If a test is performed, the set of inferences allows us to draw conclusions about a subset of components. At any point in a sequence of tests, the model can be used to compute the set of remaining failure candidates. We developed a precise algorithm to look at the information content of the tests. This algorithm selects tests such that the number of tests required to isolate a fault is minimized over the set of potential failure candidates.

4. DIAGNOSIS USING THE DIAGNOSTIC INFERENCE MODEL

Fault isolation can be mathematically described as a set partition problem. Let $\mathbf{C} = (c_1, c_2, ..., c_n)$ represent the set of components. After the jth test, a fault-isolation strategy partitions $\mathbf{C}$ into two classes. $\mathbf{F}^j = \left(c_1^j, c_2^j, ..., c_m^j \right)$ is the set of components that are still failure candidates after the jth test (feasible set). $\mathbf{G}^j = \mathbf{C} - \mathbf{F}^j$ is the set of components found to be good after the jth test (infeasible set). By this structure, a strategy will have isolated the failure when $\mathbf{F}^j$ consists of a single element or an indivisible component ambiguity group.

Let $\mathbf{D}$ represent the full set of test inference relationships between components and test points. This is formulated as a matrix representation. Let $\mathbf{S}_k$ be a sequence of k tests, $(t_1, t_2, ..., t_k)$. Let $\mathbf{F}^k$ be the feasible failure candidate set associated with $\mathbf{S}_k$. We then develop an information measure, I_k^j, for each remaining (unperformed) test, t_j, which is a function of the inference relationship and the remaining candidate failure class, say, $I_k^j = f(\mathbf{D}, \mathbf{F}^k)$ (Shannon, 1948; Dretske, 1982; Quinlan, 1986; Simpson and Sheppard, 1994). The test sequence $\mathbf{S}_k$ that is derived is obtained by optimizing at each decision point. That is, the next test in the sequence is taken as the test that maximizes I_k^j for the conditions imposed by each previous test outcome and is based on an unknown current outcome. The sequence ends when adequate information is derived for fault isolation. Although this algorithm uses the greedy heuristic (i.e., does local search), it is based upon a higher order representation and has been providing performance near the theoretical optimum (Simpson and Sheppard, 1994).

5. DIAGNOSIS AND CONFLICT MANAGEMENT WITH DEMPSTER-SHAFER INFERENCE

Our approach to diagnosis uses a modification of Dempster-Shafer (Dempster, 1968; Shafer, 1976) statistical inference in its inference engine (Simpson and Sheppard, 1994). To summarize, we compute values for two extremes of a credibility interval for every conclusion in the model. These extremes are called *Support*, s_{c_i}, and *Plausibility*, p_{c_i}, and for a given conclusion, c_i, $s_{c_i} \leq \Pr(c_i) \leq p_{c_i}$. To compute these measures, we begin with

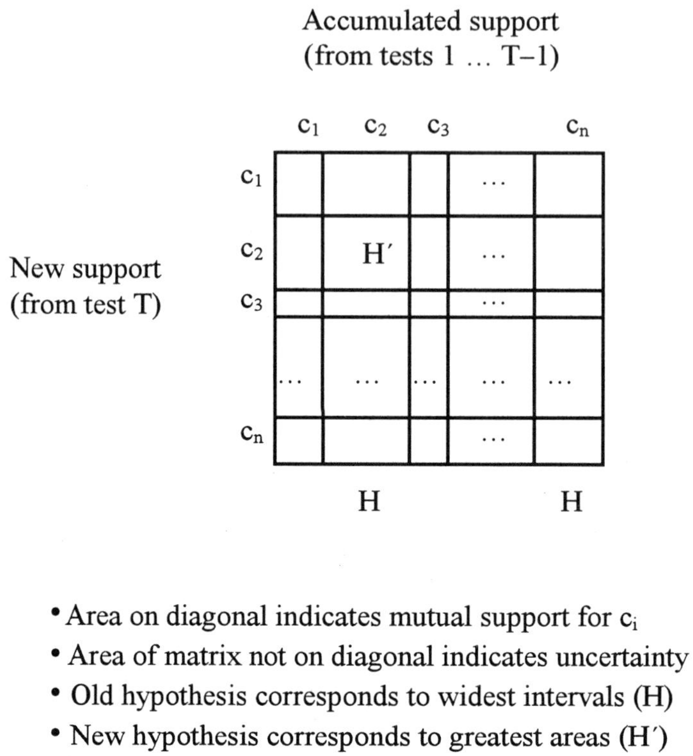

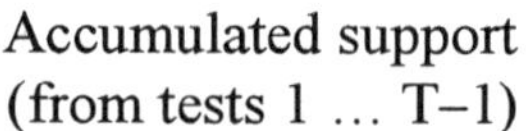

Accumulated support
(from tests 1 … T−1)
c1 c2 c3 cn
c1
c2
c3
cn
H′
New support
(from test T)
H H
• Area on diagonal indicates mutual support for ci
• Area of matrix not on diagonal indicates uncertainty
• Old hypothesis corresponds to widest intervals (H)
• New hypothesis corresponds to greatest areas (H′)

Figure 1. Dempster's Rule of Combinations

assigning a confidence value to a particular test outcome, cf_{t_j}. In our formulation, we uniformly distribute the support over all conclusions supported and apply the full weight of denial (the complement of plausibility) to all conclusions denied. Thus,

$$s_{c_i} = \frac{cf_{t_j}}{|\mathbf{C}_s|}$$
$$d_{c_i} = cf_{t_j}$$

where $\mathbf{C}_s$ is the set of conclusions supported by the evidence given in t_j, and d_{c_i} is the *denial* of conclusion c_i.

From these, we compute support and plausibility measures incrementally. We determine how much the new evidence conflicts with previously accumulated evidence (initially assuming no disagreement). Then we revise the support for each conclusion using a variant on Dempster's Rule of Combinations (Dempster, 1968) which computes normalized mutual support and denial for each conclusion using the current accumulation of support and denial with the support and denial received from the most recently evaluated test (Figure 1). Finally, to determine plausibility we keep a running average of the denial obtained thus far and subtract from one. This process is implemented using the following sequence of steps:

$$k(t) = \sum_i \sum_j \delta_{ij} s_{c_i}(t) \widetilde{s}_{c_j}(t-1)$$

$$\hat{s}_{c_i}(t) = \frac{s_{c_i}(t)(\widetilde{s}_{c_i}(t-1) + u(t)) + \widetilde{s}_{c_i}(t-1)(1 - cf_{t_j}(t))}{1 - k(t)}$$

$$u(t) = u(t-1)\frac{1 - cf_{t_j}(t)}{1 - k(t)}$$

$$\widetilde{d}_{c_i}(t) = \widetilde{d}_{c_i}(t-1) + d_{c_i}(t)$$

$$p_{c_i}(t) = 1 - \frac{\widetilde{d}_{c_i}(t)}{t}$$

where,

$$\delta_{ij} = \begin{cases} 0; i = j \\ 1; i \neq j \end{cases}.$$

A modification to the Dempster-Shafer process includes defining the *unanticipated result* (Simpson and Sheppard, 1994). This special conclusion helps to compensate for declining uncertainty in the face of conflict. The support for an unanticipated result (representing conflict) is computed whenever evidence denies the current hypothesis. For this to occur, the evidence must deny *all* of the conclusions in the hypothesis set $\mathbf{H} \in \mathbf{C}^+$ (a non-empty set of conclusions). The amount of conflict is apportioned over the number of tests executed so far, so

$$\widetilde{s}_u(t) = \widetilde{s}_u(t-1) + \frac{\chi(t)k(t)cf_{t_j}(t)}{t}$$

where χ is the number of times a conflict has occurred. When no conflict exists, support for the unanticipated result decays according to

$$\widetilde{s}_u(t) = \widetilde{s}_u(t-1)\frac{t-1}{t}.$$

Now we are ready to compute the final support measure. First note that plausibility is computed as a function of normalized denial. At each step, support is normalized as follows.

$$\widetilde{s}_{c_i}(t) = \frac{\hat{s}_{c_i}(t)(1 - \widetilde{s}_u(t))}{u(t) + \sum_{\forall c \in C} \hat{s}_c(t)} .$$

The primary computational burden of this procedure lies in determining the normalization constant of Dempster's rule. This normalizer requires a summation over all pairwise combinations of support values. It has a complexity of $O(n^2)$, where n is the number of conclusions. The calculations for combining support and denial and for computing conflict are relatively simple, being of complexity $O(1)$, and the final calculation for normalizing support is $O(n)$. Thus, the overall computational complexity of this process is $O(n^2)$ in each step.

6. DIAGNOSIS AND CONFLICT MANAGEMENT WITH CERTAINTY FACTORS

Because of this strong dependence of the support value on previously normalized data, the Dempster-Shafer calculations exhibit a temporal-recency effect. In other words, more recent events have a greater impact on the evidential calculation than more distant events. The significance of this is that the evidential statistics are not temporally independent. As a result, if the same set of tests are analyzed with the same outcomes and the same confidences but in different orders, the resulting Dempster-Shafer statistics will be different.

Because of this undesirable property, we began to explore alternative approaches to reasoning under uncertainty in which we could base our inferences on the diagnostic inference model, assign confidences to test outcomes, and perform consistent inference independent of any temporal ordering. As a guide, we began by listing several characteristics we felt were reasonable for any uncertainty-based inference system. These characteristics included the following:

- We should be able to track levels of support and denial for each conclusion in the model.
- We should be able to convert these support and denial measures to an estimate of probability given the evidence, i.e., $\Pr(c_i|e)$ that is both reasonable and intuitive.
- We should be able to apply test results in any order and yield the same result.
- We should be able to evaluate levels of conflict in the inference process, and all measures associated with conflict should have the same properties of any other conclusion in the model.

From these "requirements," we started to derive a simplified approach to reasoning with uncertain test data and discovered that we had re-derived a relatively old method called *certainty factors*. Certainty factors were first used by Edward Shortliffe in his *MYCIN* system, developed in the early 1970s and provided an intuitive approach to reasoning under uncertainty in rule-based systems that had several roots in probability theory (Shortliffe, 1976). As we started to work with certainty factors, we found they satisfied all of our requirements except for the handling of conflict. The following describes our implementation of certainty factors for the diagnostic inference model, including the creation of a conflict-management strategy that satisfies the above requirements.

As with Dempster-Shafer, we begin by noting that test outcomes either support or deny conclusions in our conclusion space. The first variation on Dempster-Shafer, is that we assign the full confidence value to all conclusions either supported or denied rather than apportioning confidence to the supported conclusions. Using the notation developed above for Dempster-Shafer, we have,

$$s_{c_i} = cf_{t_j}$$
$$d_{c_i} = cf_{t_j}$$

Obviously, as before, support is only applied to a conclusion if the test outcome actually supports that conclusion, and denial is only applied if the test outcome actually denies the conclusion.

Updating support and denial over time is straightforward and has similarities to combining probabilities. In particular, we can update support and denial as follows:

$$\tilde{s}_{c_i}(t) = \tilde{s}_{c_i}(t-1) + s_{c_i}(t) - \tilde{s}_{c_i}(t-1)s_{c_i}(t)$$
$$\tilde{d}_{c_i}(t) = \tilde{d}_{c_i}(t-1) + d_{c_i}(t) - \tilde{d}_{c_i}(t-1)d_{c_i}(t)$$

According to Shortliffe, the certainty in a conclusion is given by

$$cert_{c_i}(t) = \tilde{s}_{c_i}(t) - \tilde{d}_{c_i}(t).$$

This is not quite enough for us since $cert_{c_i} \in [-1,1]$. First we need to rescale the value such that $cert'_{c_i} \in [0,1]$. We accomplish this as follows:

$$cert'_{c_i} = \tfrac{1}{2}(cert_{c_i} + 1).$$

Then we compute the probability as

$$\Pr(c_i|e) = \frac{cert'_{c_i}}{\displaystyle\sum_{\forall c \in \mathbf{C} \cup (unt)} cert_c} \cdot$$

Note this equation includes *unt*, i.e., the unanticipated result. Recall that all test outcomes support some conclusions and deny other conclusions. Prior to doing any diagnosis, we can determine the support sets for each of the tests. Determining the denial set is done by taking the complement of the support set which adds no new information to our calculation. Further, the impact of denial is based on a single failure assumption which makes determining conflict based on denial questionable.

For any given test, we can determine the test's support set when the test passes and when the test fails. We want to compare these support sets to the support sets of other tests. In particular, for a sequence of tests, we are interested in determining the relative conflict between all pairs of tests in that sequence. Support and denial for conflict then consist of combining support and denial at each step in the sequence using the combination procedures described above. All we need now is a way to determine s_u and d_u (Sheppard, 1996).

Consider two tests t_i and t_j. These two tests may conflict in any of four possible situations—when both tests pass, when both tests fail, when t_i passes and t_j fails, and when t_i fails and t_j passes. Without loss of generality, suppose both tests fail. If we consider the intersection of the tests' support sets given they fail, we claim that if the intersection is the empty set, these two outcomes are inherently conflicting, i.e., they support completely different sets of conclusions and, in fact, deny each other's sets of conclusions (Figure 2). In this scenario, we can determine the relative amount of conflict as follows:

$$\chi(val(t_i) = \mathrm{FAIL} \wedge val(t_j) = \mathrm{FAIL}) = 1 - \frac{\left| \mathbf{C}^{\mathrm{F}}_{t_i} \cap \mathbf{C}^{\mathrm{F}}_{t_j} \right|}{\left| \mathbf{C}^{\mathrm{F}}_{t_i} \cup \mathbf{C}^{\mathrm{F}}_{t_j} \right|}$$

where $\mathbf{C}^{\mathrm{F}}_{t_i}$ is the set of conclusions supported by t_i failing.

Similarly, we can determine the relative amount of conflict *denial* associated with a pair of test outcomes. If the intersection of the support sets is not empty, then there exists a set of conclusions mutually supported by these two test outcomes. This area of mutual support indicates that the test outcomes are inherently non-conflicting, thus indicating we can deny the presence of conflict in the diagnostic process. Therefore, we can compute the relative denial of conflict between two test outcomes as follows:

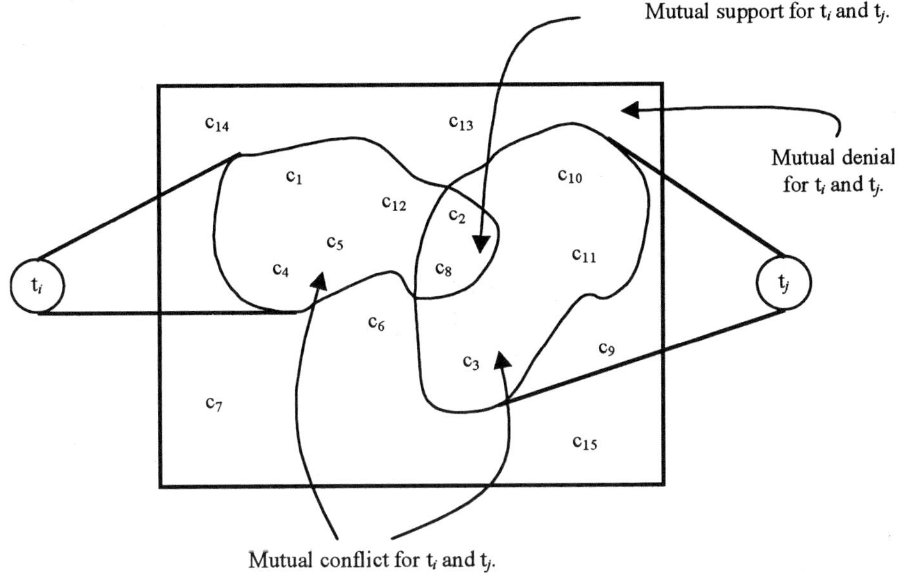

Figure 2. Determining support and denial for unanticipated result.

$$\overline{\chi}(val(t_i) = \text{FAIL} \wedge val(t_j) = \text{FAIL}) = \frac{|\mathbf{C}_{t_i}^{\text{F}} \cap \mathbf{C}_{t_j}^{\text{F}}|}{|\mathbf{C}_{t_i}^{\text{F}} \cup \mathbf{C}_{t_j}^{\text{F}}|}.$$

Individual values for s_u or d_u depend on the confidence in the test outcomes and can be computed as

$$s_u(val(t_i) \wedge val(t_j)) = cf_{t_i} cf_{t_j} \chi(val(t_i) \wedge val(t_j))$$
$$d_u(val(t_i) \wedge val(t_j)) = cf_{t_i} cf_{t_j} \overline{\chi}(val(t_i) \wedge val(t_j))$$

As tests are evaluated, we accumulate support or denial for the unanticipated result in a similar fashion to combining standard support and denial, except that a single test outcome can cause several new "events" to be added. Formally, we perform the accumulation as follows:

$$\widetilde{s}_u(\tau) = \left(\bigoplus_{i=1}^{\tau-1} s_u(val(t_i) \wedge val(t_\tau)) \right) \oplus \widetilde{s}_u(\tau - 1)$$

$$\widetilde{d}_u(\tau) = \left(\bigoplus_{i=1}^{\tau-1} d_u(val(t_i) \wedge val(t_\tau)) \right) \oplus \widetilde{d}_u(\tau - 1)$$

where $\oplus$ denotes combination as defined previously. Thus, we have to combine conflict at each step by considering all past steps in the test process. This is the most computationally expensive part of the certainty factor approach: It requires $O(T^2)$ time, where T is the number of tests performed. The complexity of computing relative conflict is $O(m^2)$ for each of the four alternatives, where m is the number of tests in the model; however, this process need be performed only once for each model.

The primary advantages to using certainty factors rather than Dempster-Shafer include reduced computational complexity and sequence independence in determining support and denial for each of the conclusions. Dempster-Shafer's primary advantage is a firmer grounding in probability theory and a larger base of practical experience demonstrating acceptable behavior.

7. INTERPRETING CONFLICT

The previous sections provided algorithms for system diagnosis in the presence of uncertainty and conflict. As we discussed in the introduction, drawing conclusions from uncertain, but consistent test information is relatively straightforward. In this section, we will focus on the problem of interpreting conflicting test results. We begin by pointing out some basic assumptions for making a diagnosis in which conflict might arise.

First, we will limit our discussion to interpreting test results that either pass or fail. Note that this is not really a limiting assumption since all tests can be reduced to a set of binary outcome tests. Further, the algorithms we provided earlier will work with multiple-outcome tests as well as binary tests. Second, we assume that the diagnostic system is focusing on identifying a single fault. This assumption will appear, initially, to be very restrictive; however, we will see that this assumption will be useful for interpreting conflict as an indicator of a multiple fault which, in turn, facilitates multiple fault diagnosis. Normal assumptions in papers on system diagnosis include that the model is correct and the test results (or at least their confidences) are correct. Since test error and model error are prime causes of conflict, we will not make these assumptions.

We believe that only three fundamental reasons exist that might lead to conflict.

1. An error occurred in testing or test recording
2. Multiple faults exist in the system.
3. An error exists in the knowledgebase or model

By providing a separate conclusion in the model for conflicting information, we provide a powerful mechanism for identifying if one of three situations exists. We point out, however, that independent analysis may be required to distinguish these potential sources of conflict in any particular instance. The following paragraphs describe approaches that have been used in actual diagnosis to identify causes of conflict.

7.1   Identifying Errors in Testing

In real systems, testing is rarely perfect. In software, it is almost impossible to keep track of system state or to track all of the ways values might be affected. Hardware systems are frequently subject to noise and other random events making interpretation of test results difficult, thus lowering confidence in what the tests are indicating. Digital testing that may not be as susceptible to noise problems still must contend with the effects of state. Finally, modern systems heavily depend on both hardware and software, and the interactions between hardware and software further compound the problem of managing test error.

A common approach to minimizing the probability of error in testing is to apply automatic test methods to the system. The thought is that automatic testing takes much of the uncertainty out of testing since the same test is applied every time, and a machine performs and evaluates the results of the test. Unfortunately, automatic testing, while eliminating many sources of error, introduce many more sources. Software must be written to run on the tester, and that software must also be tested. Until recently, much of the test software has been written from scratch with every new system to be tested leading to a high development cost and a high likelihood of repeatedly making the same mistakes in writing the tests. Finally, instrumentation used to apply test stimuli (or inputs) and interpret the response (or outputs) have physical limitations that can introduce error as well (Dill, 1995).

For example, suppose we are measuring a value in a system that may be subject to noise. In particular, the actual value (assuming nominal) may fit a normal distribution (assuming Gaussian noise), and the nominal value should appear within a specified range. (By the Central Limit Theorem, the Gaussian assumption will be valid when considering a large number of distributed variables, even if the individual variables are not Gaussian.)This situation is illustrated in Figure 3. Just as the system may have error in the output, the instrument measuring the output may also have error (based on the accuracy or precision of the instrument, among other factors). Therefore, we can model the nominal range for the instrument in a similar fashion (Figure 4). The problem arises when we overlay these curves. For example, is the value measured in Figure 5 nominal or not?

If the actual value falls on the inside of the nominal range, but we declare that the test fails, we have introduced a *false alarm* into our test system. On the other hand, if the actual value falls outside of the nominal range and we declare that the test passed, we have introduced a *false assurance* into our test system. In statistics, these errors are referred to as

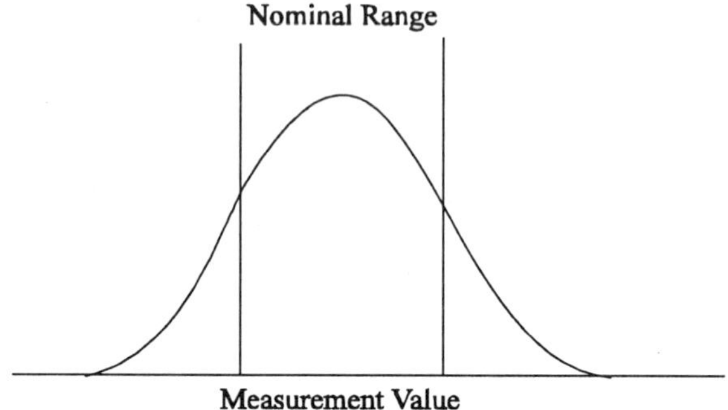

Figure 3. Nominal range for value.

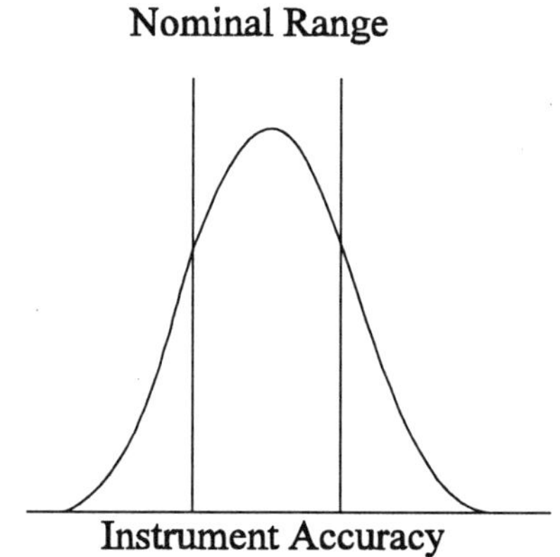

Figure 4. Nominal range for instrument.

Type I and Type II errors. Which is which depends on whether our hypothesis is that the system nominal or faulty.

Sensitivity to Type I and Type II errors depends on the variance of the noise associated with a measurement and the variance on the measurement device itself. The larger the variance, the higher the probability of test error. When such sensitivity exists, several techniques are available to reduce the likelihood of error. One approach is to modify the decision boundaries (i.e., the tolerances) on the measurement. Unfortunately, tightening the bounds can lead to additional false alarms, and opening the bounds can lead to additional false assurance. A second approach is to take multiple measurements and either vote or take an average of the measurements. This approach increases the probability of an accurate

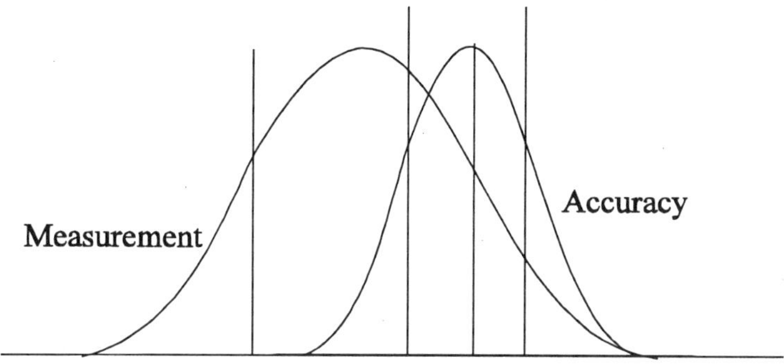

Figure 5. Uncertain pass/fail criteria.

measurement and interpretation as the number of measurements increases but can have a severe impact on testing efficiency. The approach also assumes that each measurement is independent of each other measurement, which is not always the case. Finally, one can take multiple, independent measurements at various points in the system and use the measurements as a consistency check on the other measurements.

In all three cases, when error exists in comparing the measured value to the test limits, the potential for conflict in diagnosis increases. Inconsistent test results lead to conflict or to an incorrect diagnosis. We prefer using the third approach to handling test error since it provides a means for identifying the conflict and for gathering supporting or denying evidence through additional testing. For example, we recently fielded a diagnostic system for testing the radar on an airplane (Gartner and Sheppard, 1996). This radar used built-in test (BIT) which included a set of "accumulated" BIT tests. Accumulated BIT is analogous to the repeat polling methodology for limiting test error and consists of multiple evaluations of a test during flight. If the test fails, a counter in the BIT matrix is incremented. If the count exceeds some threshold, then the test failed; otherwise, it passed.

The diagnostic system we fielded used the Dempster-Shafer methodology for managing uncertainty and conflict. A preferred approach to handling the accumulated BIT would have been to treat each instance as a separate test with a low confidence. As the test is repeated, the confidence would accumulate with the counter and provide a more accurate indication of pass or fail during diagnosis. In this case, however, the diagnostic system only saw whether or not the threshold was exceeded, and a (perhaps inappropriately) high confidence was associated with the result. When the count is near the threshold, we might see the situation depicted in Figure 5 in which the support for unanticipated result rises.

As indicated earlier, determining the cause of conflict frequently requires *post mortem* analysis of the test results to determine the cause. In the absence of additional information, it is almost impossible to determine if the source of conflict was test error. In the case described above, test error becomes likely when considering the nature of the test itself. Since a "hard" threshold is applied to the test interpretation rather than applying either a "fuzzy" threshold or a confidence based on proximity to the threshold, the chances of not detecting a fault or reacting to false alarms increases. Knowledge of the nature of the test and the way the test results are fed to the diagnostic system facilitate this analysis.

7.2   Identifying Multiple Faults

The next step in managing conflict involves identifying the possibility of multiple faults being present in the system being tested. Multiple fault diagnosis is highly complex (deKleer, 1987; Sheppard and Simpson, 1994; Shakeri *et al.*, 1995). While many diagnostic systems exist that claim to diagnose multiple faults, and indeed it is possible to estimate most likely optimal faults fairly efficiently, the ability to correctly identify a multiple fault every time in a reasonable period of time is virtually impossible.

We use the conflict conclusion (i.e., *unanticipated result*) as an indication that a multiple fault may be present. If diagnosis leads to a single fault conclusion with no conflict, the probability is high that the single fault identified is correct (ignoring the problem of masked root cause faults and false failure indications). Therefore, if testing yields inconsistent results, one possible high probability cause is the presence of multiple faults.

Typically, multiple faults are identified and addressed in single fault systems using a sequential approach to diagnosis and repair (Shakeri *et al.*, 1995). In particular, first a single fault is identified and repaired. After the fault is removed from the system, the system undergoes additional testing to determine if another fault exists in the system. If so, it is identified and repaired as before. This process continues until all faults are identified and repaired.

This approach will work (albeit inefficiently) for all multiple faults in the system, with the exception of two. The first multiple fault situation in which "peeling" does not adequately identify and repair the system is when the *root cause* of system failure is masked by another, secondary fault caused by that root cause. When the root cause fault is masked by the secondary fault, both single fault isolation and any multiple fault isolation will identify only the secondary fault. This is a direct result of the masking effect. This in turn leads to ineffective repair since repairing the secondary fault is futile.

The secondary fault recurs when the system is reinitialized because of the root cause fault.

The second multiple fault situation in which "peeling" does not adequately identify and repair the system is when a *false failure indication* occurs. A false failure indication occurs when the symptoms of two or more faults are identical to another single fault. Once again, this leads to ineffective repair of the system since repairing the indicated single fault has no effect and the system is not restored to operational status. In other words, diagnosis will not have identified the faulty components, and maintenance leaves the system in the original failed state. Multiple failure diagnosis does not eliminate the problem, but the situation is improved. In particular, we find that with multiple failure diagnosis, we will be able to identify both the single fault that would be falsely indicated under single fault diagnosis, and the multiple fault. Unfortunately, we are unable to tell the difference between the two, and the two conclusions are considered to be *ambiguous*.

Generally, root cause situations can be identified through engineering analysis prior to fielding a diagnostic system, and common false failure indications can be tied to root cause situations. Therefore, it is still possible for single fault diagnosis, tied to good repair procedures, to address these problems. But what happens if multiple, independent faults exist in the system and we do not have the luxury of the time required to use the peeling strategy? Alternatively, what if certain faults simply cannot be removed from the system (e.g., a fault on a satellite in orbit)?

Conflict occurs when test results either directly or indirectly contradict each other. This is manifest in Dempster-Shafer when a test result denies the current hypothesis and in certainty factors when the support sets for two tests are disjoint. When conflict occurs, one possible cause is the presence of two independent multiple faults. Typically, when examining the ranked list of faults returned by these two techniques, the technician treats them in ambiguity (i.e., they look for one fault among the list). When conflict occurs, this can be used as a flag to the technician that more than one of the faults in the list may be present. Generally, the technician considers the faults in order of probability. This can still be used. In addition, with the indication of conflict, the diagnostic system can apply a separate process among the top-ranked faults to determine if their combined symptoms match the observed symptoms. If so, this is a strong indication that the multiple fault is present.

The radar system described previously provided a simple example where a multiple fault appears to have occurred (Gartner and Sheppard, 1996). In this case, the multiple fault was benign in that both faults were contained within the same replaccable unit, and the maintenance personnel were only interested in fault isolating to the replaceable unit. Personnel at the next level of maintenance would have a problem when they attempted to repair further.

In this case, the problem was the presence of two antenna faults. Two independent BIT codes were obtained directly indicting two separate failure modes of the antenna. The diagnostic system noted the conflict and called for a separate "initiated" BIT (IBIT) to be run. The result was identification of the antenna with both faults identified and equally supported in the fault candidate list.

7.3   Identifying Modeling Errors

Modeling systems for diagnosis frequently entails capturing knowledge similar to traditional knowledge engineering tasks in expert system development. One common complaint with expert systems is that they are brittle, that is they are frequently incapable of handling situations in which the information obtained is inconsistent with the knowledge base. The availability of a special conclusion representing the presence of such an inconsistency, and the ability to further process the model in light of the inconsistency provides tremendous power in overcoming the brittleness of the traditional expert system.

If the test results are understood and no multiple faults exist in the system, then conflict may be caused by error in the model. The process of diagnostic modeling is extremely difficult and error prone. Identifying errors in the model is equally difficult. So far, the best approaches found for model verification have been based on analyzing logic characteristics of the models and performing fault insertion in the diagnostic strategies. Fault insertion is generally the most effective approach but can be time consuming and damaging to actual systems. Using fault insertion together with fault simulation eliminates the problem of potentially damaging the system; however, the verification problem now includes verifying that the simulation model is correct. Otherwise, the only assurance gained from fault insertion in the simulation model is that the diagnostic model represents the simulation, not the actual system.

One of the advantages to modeling systems with the diagnostic inference model is that logical characteristics of the model can be extremely useful in verifying the model. Identifying ambiguity groups, redundant and excess tests, logical circularities, masked and masking faults, and potential false failure indications can all indicate potential problems with the model. Details on using these characteristics in verification are provided in Simpson and Sheppard (1994).

Even with the best verification tools and techniques, errors invariably remain (especially in complex models). During actual testing, it is frequently difficult to identify when model errors arise. However, with the *unanticipated result* included in the set of possible conclusions, a "flag" can be raised whenever a model error might have been encountered. As with test error and multiple fault diagnosis, if the model is error, it is likely at some point for a test result to be contrary to what is expected—in which case a conflict occurs and the *unanticipated result* gains support.

In the case of the radar system, we were able to find model errors since two types of models were available for comparison—a system-level model and multiple replaceable unit-level models (Gartner and Sheppard, 1996). (While the two types of models were not derived independently, they offer alternative representations of the system.) When test results were processed through the model representing the transmitter, no conflict was encountered and a large ambiguity group consisting of wiring, the transmitter itself, and the computer power supply. When the same test results were run through the system-level model (the model of choice in the field since the replaceable unit did not need to be known prior to testing), the same ambiguity group was concluded as the most likely fault, but considerable conflict was encountered during testing. No conflict was encountered in the replaceable unit model.

When we noticed this discrepancy, we looked at the diagnostic log files for the two runs. In the case of the replaceable unit-level model, only one test was considered, and that test pointed directly at the fault in question. At the system level, seven tests were considered (including the one from the replaceable unit level). Four of these tests did not appear in the replaceable-unit level model. Six tests, included these four, depended on the common test in system-level model and all passed where the common test failed. Since each of these tests depended on the common test, when the common test failed, each of these additional tests should have failed as well. Since they did not, they were in conflict with previous experience.

This discrepancy led us to take a closer look at the tests in question. One of the conflicting tests that appeared in both models served as a "collector" of test information in the BIT. In other words, all test information was funneled through that test such that if any other test failed, that test was expected to fail as well. But this test had another odd characteristic. Whenever certain of these other tests failed, this test was disabled and not run. The diagnostic system assumed that a test that was not run would have passed (a bad assumption) and this led to the conflict. The failing test determined whether the transmitter was transmitting by examining the least significant bit in a data word sent in a loopback test.

The more appropriate representation of this situation in the model would have used what is referred to as a "linked test outcome." Traditional diagnostic inference models assume "like" inferences. In other words, if a test fails, it is assumed test inferences drawn from that test will consist only of failing and if the test passes, test inferences drawn from that test will consist only of passing. But this is not a realistic assumption, so the diagnostic inference model permits declaration of special inference linkages—linked test outcomes—in which a passing test can result in other tests being inferred fail or unavailable and a failing test can result in other tests being inferred pass or unavailable. The skipped test in the radar transmitter model should have been linked to the each of the tests that would have caused it to be skipped such that failure of those tests would cause an inference of the skipped test being unavailable. This was not used in the model resulting in an error. In fact, it was found that a "hard" dependency existed between the collector test (the skipped test) and the communications test in the system model and *no* dependency existed at all in the replaceable unit-level test. Both models were in error but for different reasons.

8.   SUMMARY

The process of diagnostic modeling and diagnostic testing is complex and often leads to conflicting information. Traditional tree-based approaches may miss apparent conflicts and provide reduced diagnostic accuracy. Whether interpreting uncertain test results, handling multiple faults, or contending with modeling error, the diagnostics used should support gathering information capable of identifying conflict and assessing the cause of the conflict. We presented two approaches for identifying conflict within the framework of the diagnostic inference model and provided several examples of how one can use the identification of conflict to improve the diagnostics of the system.

Uncertain and inexact testing can lead to conflict in diagnosis in that the outcome of the test as measured may not accurately reflect the system being tested. This can occur when the error range of the test instrument compared to the tolerance of the expected test results overlap significantly. Identifying when this situation exists involves a detailed understanding of the failure modes detected by the test and the error range of the instrument relative to the test. The occurrence of conflict in testing can indicate the need for closer examination of the test results by identifying that unexpected test results have occurred.

In addition to the ability to raise the concern of problem tests, detecting and analyzing the presence of conflict can be useful in multiple fault diagnosis when the diagnostic procedure begins with the assumption of a single fault. When multiple faults exist, test results can be consistent with one of the faults but inconsistent with the other. Since the process of multiple fault diagnosis is computationally complex, using the presence of conflict to flag the need for checking for multiple faults is a reasonable and computationally efficient approach to take.

Finally, since modeling and knowledge engineering is a complex and error prone task, the diagnostics can be used to assist modeling by identifying when inconsistent or illogical conditions arise. Such inconsistency is identified in the diagnostics with the *unanticipated result* gaining support. If the certainty in the test results is correctly represented and no multiple fault exists in the system, then an analyst can assume an inadequacy exists in the model. Test outcomes can result in a wide range of possible inferences, and unexpected inferences can be identified with conflict. By examining the tests evaluated and the conclusions drawn, an analyst can localize the potential cause of the conflict and identify possible problems in the model.

To date, little discussion has occurred on the positive role of conflict in system test and diagnosis. Conflict has always been regarded as something to be avoided. But conflict can provide valuable information about the tests, the diagnostics, and the system being tested. In this chapter, we attempted to describe an approach for capturing and quantifying the amount of conflict encountered in testing and to describe approaches to using the conflict to benefit the diagnostic process, thus leading to more robust overall system diagnostics.

9.   ACKNOWLEDGMENTS

This chapter describes the results of work that has been performed over several years, and the authors have received input and guidance from several people to improve the techniques and the chapter itself. We would like to thank Brian Kelley, John Agre, Tim McDermott, Jerry Graham, Don Gartner, and Steve Hutti for their comments as the algorithms were developed and the system fielded.

10.   REFERENCES

Cantone, R. And P. Caserta. 1988. "Evaluating the Economical Impact of Expert System Fault Diagnosis Systems: The I-CAT Experience," *Proceedings of the 3rd IEEE International Symposium on Intelligent Control*, Los Alamitos, California: IEEE Computer Society Press.

Davis, R. 1984. "Diagnostic Reasoning Based on Structure and Behavior," *Artificial Intelligence*, 24:347–410.

deKleer, J. 1987. "Diagnosing Multiple Faults," *Artificial Intelligence*, 28:163–196.

Dempster, A. P. 1968. "A Generalization of Bayesian Inference," *Journal of the Royal Statistical Society*, Series B, pp. 205–247.

Dill, H. 1994. "Diagnostic Inference Model Error Sources," *Proceedings of AUTOTESTCON*, New York: IEEE Press, pp. 391–397.

Dretske, F. I. 1982. *Knowledge and the Flow of Information*, Cambridge, Massachusetts: The MIT Press.

Gartner, D. and J. Sheppard. 1996. "An Experiment in Encapsulation in System Diagnosis," *Proceedings of AUTOTESTCON*, New York: IEEE Press, pp. 468–472.

Pearl, J. 1988. *Probabilistic Reasoning in Intelligent Systems*, San Mateo, California: Morgan Kaufmann Publishers.

Peng, Y. and J. Reggia. 1990. *Abductive Inference Models for Diagnostic Problem Solving*, New York: Springer-Verlag.

Pople, H. E. 1977. "The Formulation of Composite Hypotheses in Diagnostic Problem Solving: An Exercise in Synthetic Reasoning," *Proceedings of the 5th International Conference on Artificial Intelligence*, pp. 1030–1037.

Quinlan, J. R. 1986. "The Induction of Decision Trees," *Machine Learning*, Vol. 1, pp. 81–106.

Shafer, G. 1976. *A Mathematical Theory of Evidence*, Princeton, New Jersey: Princeton University Press.

Shakeri, M., K. R. Pattipati, V. Raghavan, A. Patterson-Hine, and T. Kell. 1995. "Sequential Test Strategies for Multiple Fault Isolation," *Proceedings of AUTOTESTCON*, New York: IEEE Press, pp. 512–527.

Shannon, C. E. 1948. "A Mathematical Theory of Communications," *Bell Systems Technical Journal*, Vol. 27, pp. 379–423.

Sheppard, J. W. 1996. "Maintaining Diagnostic Truth with Information Flow Models," *Proceedings of AUTOTESTCON*, New York: IEEE Press.

Sheppard, J. W. and W. R. Simpson, 1994. "Multiple Failure Diagnosis," *Proceedings of AUTOTESTCON*, New York: IEEE Press, pp. 381–389.

Shortliffe, E. H. 1976. *Computer Based Medical Consultations: MYCIN*, New York: American Elsevier.

Simpson, W. R. and J. W. Sheppard. 1994. *System Test and Diagnosis*, Norwell, Massachusetts: Kluwer Academic Publishers.

Wilmering, T. J. 1992. "AutoTEST: A Second-Generation Expert System Approach to Testability Analysis," *Proceedings of the ATE and Instrumentation Conference West*, pp. 141–152.

Chapter 7

System Level Test Process Characterization and Improvement

Des Farren
Motorola

Wai Chan
Digital Equipment Corporation

Anthony P. Ambler
University of Texas at Austin

Keywords: Test economics, system level test, event rate modeling, NHPP modeling, process optimization.

Abstract: The implementation of test improvements can have a significant impact on product lifecycle costs. However, the time lag in accruing potential benefits can be problematic when attempting to justify the required investment. This can be overcome if a phased implementation is adopted where each interim level of investment delivers less substantial but more immediate gains. This chapter demonstrates that techniques previously used to investigate cost-saving opportunities at strategic and lifecycle level can be used within a system manufacturing process to achieve tangible short-term benefits. The failure profile of a system level test process is characterized and used to drive defect elimination and optimum test times. Various "stopping criteria", product-process comparisons and improvements are discussed.

1.   INTRODUCTION

Investigation of the cost-effectiveness of system level testing was initiated some years ago because, like semiconductor and PCB processes previously, it was experiencing increasing cost pressures. From the beginning it was clear that a lifecycle view was essential but very few established characterization methods were available. An approach called Event Rate Analysis was developed as the kernel of our cost models and these showed that significant savings could be achieved. They also confirmed that the lowest costs generally corresponded to maximum customer quality. As with all lifecycle approaches, a fundamental problem is that savings can only be realised if everyone along the lifecycle chain sees an advantage in co-operating. With this in mind, the work described here was initiated to demonstrate the shorter-term benefits from implementing a consistent approach.

Test strategy alternatives may significantly influence lifecycle costs (Turino, 1994) and various analysis methods can help identify key lifecycle cost drivers (Dear *et al*, 1992; Dislis *et al*, 1989; Dislis *et al*, 1992; Moore, 1994; Tegethoff and Chen, 1994; Farren and Ambler, 1994; Farren and Ambler, 1995). System-level test processes present many unique challenges due mainly to the diverse nature of the defect profile (Gray, 1986; Maxwell and Aitken, 1993). An underlying characteristic in many system-level defect spectra is that of fault and error latency (Shin and Lee, 1986; Czech and Siewiorek, 1992; McGough *et al*, 1983; Chillarege and Iyer, 1987; Chillarege and Bowen, 1989). These latencies may range in time from minutes to days and must be taken into account in assessing any system-level test process, whether it is in the design, manufacturing or customer phases of the lifecycle.

In many cases, the complexity of most hardware-software systems, in addition to the variety of potential defects, prohibits the use of deterministic analytical methods. In these circumstances, a practical alternative is to adopt a statistical modeling approach to characterizing the testing process. One such method employs a suitable time-domain model to isolate the main features of the event occurrence profile during testing (Goel and Okumoto, 1979; Ohtera *et al*, 1990; Krten and Levy; 1980; Wohl, 1982). A wide range of system-level hardware test processes can be modeled using event rate analysis and previous applications focused on lifecycle cost optimization and strategy selection (Farren and Ambler, 1994; Farren and Ambler, 1995). The potential savings from these investigations can be significant but may never be realised if the methodologies are not fully implemented.

Typically, a major obstacle to successful implementation is convincing relevant design and manufacturing groups of potential benefits. This can be compounded by short-term approaches to planning and the inherent time lag in demonstrating lifecycle cost savings. Fortunately, the event-rate characterization methods underlying our lifecycle analyses also provide information that can be used to gain short-term cost savings through defect elimination and test time reduction. These tangible and immediate benefits can provide an incentive to embark on more comprehensive lifecycle modeling.

Previous investigations (Farren and Ambler, 1994; Farren and Ambler, 1995) confirmed that it is rarely cost-justifiable to compromise on outgoing quality and the analysis presented here is based on the premise that the optimum test time is one which minimizes the defect level in the field. While overall lifecycle costs are minimized by setting test times based on outgoing quality, attention quickly turns to in-house manufacturing costs. Our analysis showed that the cost of failure becomes a dominant factor, not just within manufacturing, but also at the lifecycle level where it can comprise 50% of total test and support costs.

A specific practical application of event rate analysis within a manufacturing system test environment is presented here. It compares actual and quality-optimized test times and investigates opportunities for defect elimination. The test processes are described, as is the need to specify and prepare appropriate datasets for analysis. The methods used to apply the event rate model are discussed and certain limitations are identified. The optimum test time corresponds to a point beyond which a small increment generates a relatively minor improvement in outgoing quality. The results depend on which method is employed in evaluating test "stop" times and three approaches are described. The analysis and interpretation of a number of datasets illustrates how the model characterizes the process and overall improvements are presented.

2.   THE SYSTEM TEST PROCESSES

System test forms the upper level of a test hierarchy and, even though product configurations can vary, the hardware has usually been extensively tested before reaching the system assembly stage. This lower-level testing of components and boards often consists of fixed ATE or BIST sequences and, usually, the BIST sequences can be repeated at more than one level in the system hierarchy.

Reliability testing is not usually carried out on fully integrated systems and the main objective of system-level test is to uncover latent functional faults. Repeating tests already used at lower levels is often a waste of time. Any new tests at the system level should target defects that either don't exist at lower levels or have escaped earlier tests. Random testing, in the form of a system-level exerciser, can be very effective in these circumstances. It can generate operational conditions not previously encountered and reveal subtle defects such as component interaction. However, a disadvantage of random testing is that a particular fault condition can be difficult to replicate. This means that fault isolation can be laborious and isolation accuracy can depend greatly on the fault management capability of the system under test. In general, the broad defect spectrum, the diversity of test methods and the loose definition of a system all compound the task of characterizing a system-level test process.

Our system-level processes suffered from most of these problems. They consisted of both deterministic and random sequences that formed a hierarchical test set. These were usually applied from the bottom up and experience confirmed that the system-level exercisers were effective in revealing the last few defects. Unfortunately, they were also significant test time drivers. Fault isolation at lower levels of the system test hierarchy relied on the diagnostic capability of the power-on and structural tests. Failures during the exerciser sequences were typically isolated using fault management information provided by the system. However, isolation accuracy depended heavily on the ability of production technicians to interpret this information.

In the analysis presented here, two different system level test processes were modeled. Each consisted of a set of system-level test programs executed under separate operating system environments. In both cases, these exercisers generated pseudo-random system activity. However, the target fault set was different for each process. While process α focused on kernel system hardware, process β exercised this as well as all peripheral devices and attempted to do so under typical customer load conditions.

3. DATA COLLECTION AND PREPARATION

In many system-level manufacturing test processes, the hardware reliability failure rate will be barely perceptible. However, in general, the discovery rate profile for existing, latent, faults must be segregated from the background of reliability failure occurrences before it can be characterized using the event rate model. This requires a definition of each category and the segregation of all failure events using error symptoms, failure verification and repair data. It may also be necessary to eliminate zero-hour failures from the fault discovery dataset to ensure they do not overly influence the model parameter estimation. Having obtained a "clean" set of test discovery failures, a table can be generated which lists the time-to-failure for each countable event. Our approach was to use such a data table to characterize higher-level test as a non-homogeneous Poisson process (NHPP) with a suitable intensity function. We selected a modified software reliability model that reflected the rate of occurrence of test events as a function of test time.

A broad selection of products was examined and the NHPP event rate model defined in (Goel and Okumoto, 1979), and modified in (Farren and Ambler, 1994), was applied to datasets representing the event discovery profile in the manufacturing system test processes. These datasets reflected two broadly different types of test process and most event rate graphs in the following sections show the actual data and the fitted model based on two different parameter estimation methods. It should be emphasized that we were attempting to model a fault discovery process and *not* a hardware reliability process and the reasons for choosing this function were twofold. Firstly, it represented our data very well and, secondly, it was derived from a set of assumptions that made sense from a system test viewpoint. Another important feature was that this event rate profile inherently accounts for fault isolation inaccuracies. This is because the event rate is proportional to the expected number of faults remaining in the system. If faults are not being removed, the Event Rate will remain higher for longer.

4.　　THE EVENT RATE MODEL

The event rate model used was a modified Goel-Okumoto (G-O) model where the event occurrence rate is:

$$ae^{(-bt)} + c$$

where t is test time and a, b and c are model parameters.

The complete model fitting procedure consisted of first estimating the parameters a, b and c using the least squares method and then taking these results as initial values for the maximum-likelihood estimation (MLE) procedure.

Under the modified G-O model, the expected number of events after testing for t hours is

$$EN(t) = (a/b)(1 - e^{(-bt)}) + ct$$

If we test a total of k systems for T hours and we observed N_i events at T_i, $i = 1,...,L$; $0 < T_1 < ... < T_L < T$, then the least squares method attempts to find the estimates of a, b and c that minimize the sum of $L+1$ differences

$$(D_1)^2 + ... + (D_L)^2 + (D_{L+1})^2$$

where

$$D_i = N_1 + ... + N_i - kEN(T_i)$$

for $i = 1, ..., L$, and

$$D_{L+1} = N_1 + ... + N_L - kEN(T).$$

The last term D_{L+1}, which incorporates the information that no events occurred between T_L and T, may have a significant influence on the estimates when $T\text{–}T_L$ is large. It is important to note that the least-squares method depends on the model only through the expected number of events and it is not suitable for the derivation of confidence limits.

For interval estimates of the model parameters, we turn to the method of MLE. When we test systems for a fixed duration, the observable random variables are the total number of events and the time of occurrence of the events. The MLE method attempts to find the estimates of a, b and c that maximises the likelihood function which is the joint probability density of the total events $N_1+...+N_L$ and the L occurrence times T_1, ..., T_L. Under the NHPP assumption, the random variable $N_1+...+N_L$ has a Poisson distribution and the conditional density of the occurrence times T_1, ..., T_L given that $N_1+...+N_L$ events have occurred can be determined from $EN(t)$. However, the maximum likelihood estimates for the modified G-O model cannot be expressed in closed form. The problem can only be solved numerically. We use the least squares estimates as initial values for the numerical procedure we have implemented to solve the maximisation problem. The covariance matrix of the parameters, which are needed in the confidence interval calculation, are obtained as the inverse of the information matrix.

In many cases the modeled profile from both least-squares and MLE was similar and, in those circumstances where the algorithm failed to converge on a solution, the least-squares was assumed to be a reasonable approximation. Further work is required to improve the success rate with the numerical algorithm in the MLE procedure.

5. OPTIMUM TEST TIME

Test coverage is a function of test time and the optimum test time depends on the factors being considered. Additional testing can mean additional manufacturing cost, while insufficient testing results in increased field costs. From an economic point of view, the optimum test time is that which minimizes overall costs (Farren and Ambler, 1994). From a quality standpoint, the optimum test time is that which minimizes the outgoing defect level. The latter definition is considered here and a number of manufacturing datasets are examined with this in mind.

6. METHODS OF EVALUATING STOPPING CRITERIA

We evaluated three methods of identifying the point at which further testing has an insignificant effect on the relative outgoing quality. The first method involves setting a threshold event rate somewhere above the steady-state value. We experimented with a threshold of 105% of the value of the parameter C. Method 2, the Area Ratio Test, sets the test time by evaluating defect probabilities in small incremental time periods. Method 3 determines a test time that ensures that the number of defects left in the system is below a specified level. We found that the different methods can have a significant influence on the recommended test time, depending on the estimated values of the model parameters. While we settled on Method 2 as the most consistent, all three are discussed here in more detail.

<u>Method 1</u>. Occurrence rate $= 1.05\ c$

In the MLE procedure, the likelihood function is maximised under the constraints that the parameters $a,\ b$ and c are non-negative. These constraints are consistent with the NHPP assumptions. Therefore the estimated occurrence rate $ae^{(-bt)} + c$ will be decreasing in t. The overall occurrence rate will reach 105% of the steady-state value of c at $t = -\mathrm{Log}(.05c\,/\,a)\,/\,b$. The upper 95% confidence limit of t can be used as a stopping criteria.

<u>Method 2</u>. Area ratio

The test duration determined by the first method will be highly influenced by small values of c. In this method, we note that at steady state, the probability that no event will occur in the interval $(T,\ T+\Delta T)$, is independent of the starting time T. Therefore, we can conclude that time to steady state has been reached if the ratio

$$R = \frac{P(\text{No event in}[T - \Delta T, T])}{P(\text{No event in}[T, T + \Delta T])}$$

is sufficiently close to unity. To find the probability P(No event in $[T-\Delta T, T]$), we first compute the expected number of events, which equates to the area under the event rate curve over the same time interval. Since the number of events has a Poisson distribution under the NHPP assumption, the desired probability is $e^{(-Area)}$. If we fix the length of the interval ΔT, then the ratio R will be increasing in T with a limit equal to 1. An incremental test time of ΔT hours will be deemed unnecessary if the test duration of T hours is chosen so that the ratio R is at least 0.9999. As in the first method, the upper 95% confidence limit of T can also be obtained based on the MLE of the model parameters.

<u>Method 3</u>. Relative increase in outgoing defect level

Results of the two previous methods cannot be readily quantified in terms of incremental/decremental outgoing quality. We now make use of the fact that the modified G-O process can be represented as the superposition of the Goel-Okumoto NHPP with an independent Poisson process. The event rate of the Goel-Okumoto component is $ae^{(-bt)}$ while the event rate of the Poisson component is c. The event rates are additive as a result of the independence assumption. An increased test time will have no effect on the series of events generated by the Poisson process as it is already at steady state. Therefore test time should be determined based solely on the Goel-Okumoto process. If testing is terminated after t hours, then the expected number of events remaining from the Goel-Okumoto process is $(a/b)e^{(-bt)}$. To relate it to the decremental outgoing quality, we compute the percentage of systems with no additional events occurring after t and obtain

$$e^{-(a/b)e^{(-bt)}}$$

Therefore, by stopping test at t, an additional

$$1-e^{-(a/b)e^{(-bt)}}\%$$

of the systems will experience failures. In order to affect no more than, say, 1% of the systems, we solve for t and obtain

$$t = -\mathrm{Log}(-b\mathrm{Log}(.99)/a)/b.$$

As in the previous methods, we use the upper 95% confidence limit of t as the stopping criteria.

7. IMPLICATIONS OF NON-ZERO VALUE FOR c

In finding the most appropriate representation of the event profile, the model fitting algorithm can yield a non-zero value for parameter c. Considering the original model assumption that event rate is proportional to the number of faults remaining in the system, a non-zero value is significant in a number of ways. A very low value of c could represent the background reliability failure rate but this should only become apparent with very long test times or under accelerated test conditions. A more realistic interpretation is that a test time has been reached where the process is still capable of detecting faults but these faults are no longer being successfully isolated and removed. The precise cause of the constant failure rate (CFR) component can only be determined from a detailed investigation of the process and from root-cause failure analysis. The implication, from a test time optimization viewpoint is that, once the CFR region has been reached, further testing with the same process will yield no additional improvement. Quality concerns with these "unremovable" defects must be addressed by improving diagnostic and fault isolation methods.

8.   CHARACTERIZATION RESULTS

This section discusses the results and implications of our analysis of the manufacturing test event profiles. We deployed our approach across all three dimensions of process, product and time-span. We applied it to more than 10 different types of systems, ranging from PCs and workstations to large datacentre products. We characterized five types of test processes, all of which used different system level exercisers. We also looked at the same process/product combinations at different points in time to understand how they had matured and finally, we compared datasets from two different manufacturing plants. The event rate model proved to be adequate in the majority of cases and endorsed our decision to adopt it.

Products AA, BB and DD are low-end servers or workstations, running process α, and Figures 1, 2 and 3 represent their cumulative event profiles. It can be seen that both estimates of the model are reasonable approximations of the actual dataset. However, the limited data available for Product AA resulted in the least accurate fit and there is a significant difference between the least-squares and the MLE estimates of the parameters. The confidence intervals for this product reflect this limited dataset and force more conservative parameter estimates to be used in the model. The cumulative failure graph for product FF, a mid-range server running process β, is shown in Figure 4, together with a comparison of actual and predicted event counts in the accompanying table. Again, the model is a reasonable representation of the process behavior under these conditions. The stopping criteria discussed earlier were applied to these models and appropriate test times were set for each of the products which, in all cases, were shorter than the current values.

Figure 5 shows event datasets and models for a midrange server, designated here as Product EE. Each dataset represents three months of production testing and the second analysis took place three months after the first. There were no significant product or test software changes between analyses.

Figure 1. Product AA Cumulative Event Profile

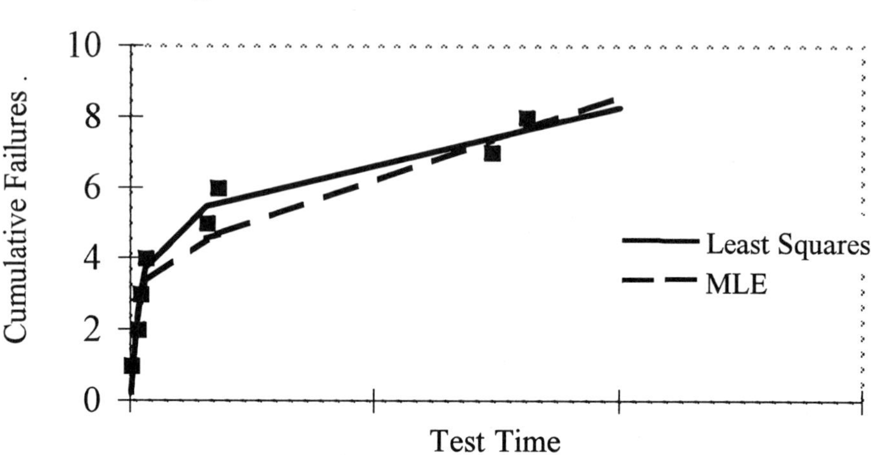

Figure 2. Product BB Cumulative Event Profile

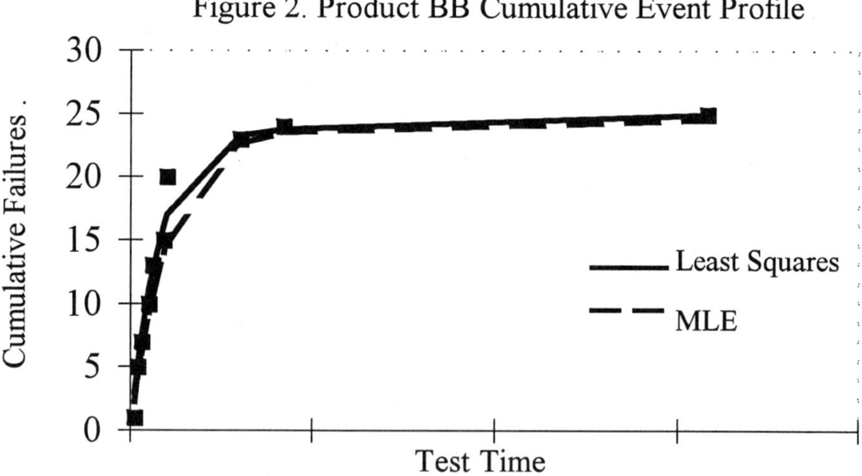

Figure 3. Product DD Cumulative Event Profile

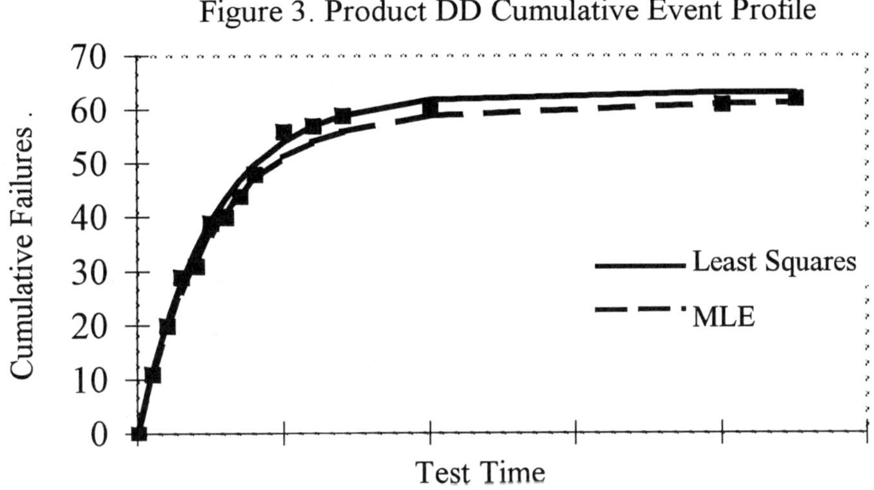

Figure 4. Product FF Cumulative Event Profile

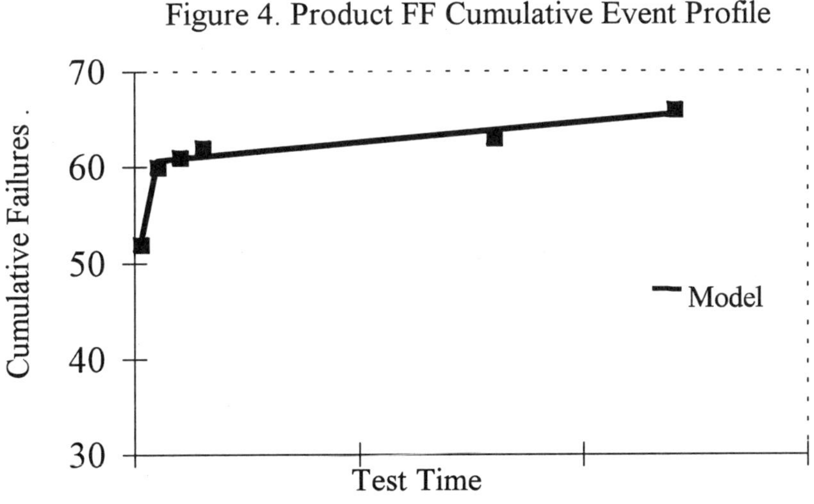

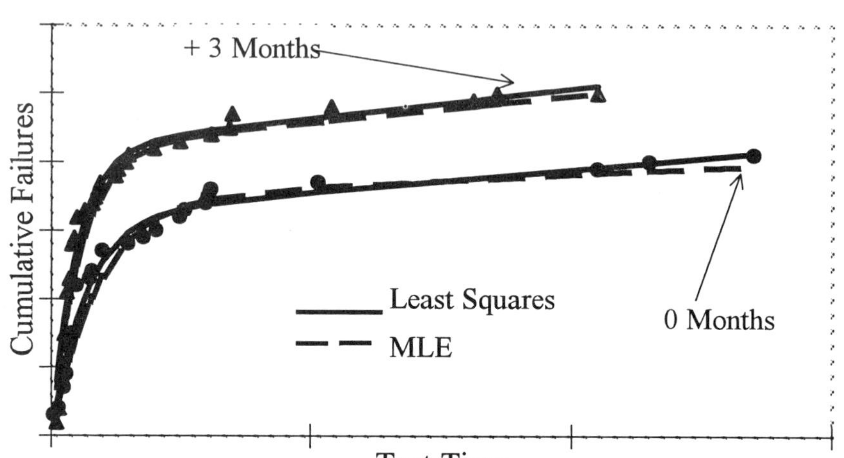

Actual versus Predicted Test Events (see Fig. 5)		
Actual	**Model**	**Error**
52.00	51.99	- 0.0%
60.00	60.65	+1.1%
61.00	60.88	- 0.2%
62.00	61.10	- 1.5%
63.00	63.84	+1.3%
66.00	65.53	- 0.7%

The most obvious difference is that the overall failure rate has increased significantly (20%) during the 3 months preceding the second analysis. This was due to a fall in the quality level of material coming into the system test process and was quickly detected by the standard quality control systems. There was some concern about whether test times should be increased to maintain outgoing quality. However, the estimated value of parameter b was almost the same in both cases (in fact it was slightly higher for the second dataset) and this allowed the previous test time and outgoing quality level to be maintained, despite the lower incoming quality. In the absence of this type of process characterization there may have been pressure to increase test time in an attempt to compensate for the poorer incoming quality.

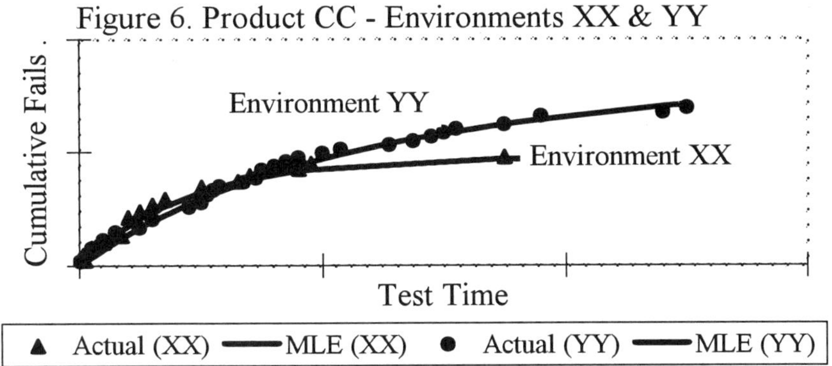

Product CC is a high-end server and the test data is shown in Figure 6. The two datasets represent the test environments, XX and YY, in two different manufacturing plants over the same time period. However, both processes tested the same basic product using identical test software. Clearly there is a difference in both the level of failures and the rate at which they are detected and this generated widely different test time recommendations. This discrepancy prompted an investigation which concluded that the difference in the level of failures was due to a difference in configuration levels. Products manufactured in environment YY contained, on average, more processor, memory and/or I/O adapters and this resulted in a higher system defect level.

The differences in the detection rate were attributed to two different aspects of the fault isolation process. One was the training and experience level of debug technicians in Environment YY and the other was a new version of the fault management software under evaluation in Environment XX. The combined effect of these was that the fault isolation accuracy in Environment YY was lower than that in Environment XX. In other words, the probability of *not* removing a defect when it was detected was higher in YY and this had the effect of sustaining a higher event rate later into the test period. This resulted in a lower estimate (50%) for parameter b in Environment YY.

Figure 7 shows the same modeled data in the form of event rate profiles rather than cumulative failure profiles. This graph highlights another significant difference in the two datasets. It can be seen that the estimate for parameter c is higher for Environment YY. This is consistent with the fault isolation difficulties and means that there is a higher level of residual, or apparently unremovable, defects in products leaving Environment YY. This adversely impacts the outgoing quality level in a manner which cannot be compensated for by increasing test time. The main outcome of this particular analysis, in addition to determining an appropriate test time for the product, was that training levels were increased and the new fault management software was put on general release. Subsequent test profile analysis showed a significant improvement in Environment YY.

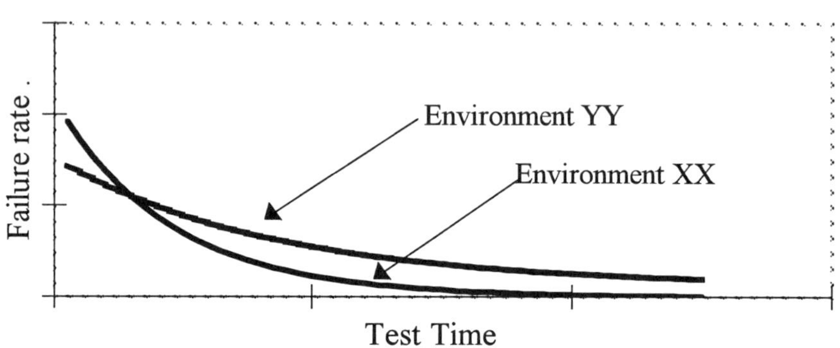

Figure 7. Product CC - Event Rate Comparison

This investigation was a good example of how the characterization methodology can lead directly to a more thorough understanding of the process behavior. Furthermore, the event rate analysis showed that the two different types of test process, and a number of different product types, can be adequately represented by the same model.

9.   PROCESS IMPROVEMENTS

The test time analysis methodology was introduced into two manufacturing sites and an overall test time reduction of 38% was achieved, with no adverse impact to customer quality. The analysis was repeated at regular intervals to ensure that the process remained tuned to changes in the defect profile. Subsequent analyses show less substantial improvement and, in some cases, slight increases were required to maintain outgoing quality. The overall reduction in test times yielded significant incremental process capacity and improved delivery predictability.

Comparisons of similar products in the different sites, as well as examination of events driving test times, yielded a number of improvements in defect rates. For example, the analysis of Product CC lead to a 35% reduction in test process events with a corresponding saving in failure costs. In one case, a detailed investigation showed that a longer test time was not warranted, despite an increase in incoming defect levels, as all new events were detected within the current period and outgoing quality was maintained.

The ability to show these improvements without impacting customer quality was viewed very positively and established the characterization method as a valuable tool. A further direct benefit of the analysis method was that the event rate graphs proved to be a valuable process management aid. They conveyed the main features of the process succinctly and, when labeled with the specific event types, immediately directed attention to appropriate areas of improvement.

10.   SOME OBSERVATIONS

In the course of implementing this analysis method, we uncovered some points worth mentioning. Our first observation was that it takes some time, and effort, to establish a new way of describing a process. Many test engineers familiar with more traditional concepts, can find a probabilistic approach a little too abstract for day-to-day work. This creates some inertia that must be taken into account in the project.

Another problem encountered was the need to constantly remind engineers that we were not modeling reliability failures. This is particularly important during data preparation, where test and reliability failures must be separated prior to analysis.

We also found that we could not rely on the model without a thorough understanding of the process. A successful approach provided the analysis capability *in support* of normal test engineering responsibilities. In this way, test engineers prepared the data, interpreted the results and made the final decisions.

Over the course of the year we realised that the ability to understand test capability lead directly to more focused goal-setting. It was possible to see where the process is today and what realistic improvements might be achievable.

11.   POSSIBLE ENHANCEMENTS

We discussed some ideas about enhancing the characterization process to make it more effective. The current analysis is very manual and we employ a number of tools including spreadsheets and customized statistical analysis software. An obvious improvement is to automate as much as possible and integrate it with the manufacturing data collection systems.

We are aware that different defect types may exhibit unique versions of the basic distribution. As a further aid to root-cause analysis we are considering adding the capability to model the profile of individual event types.

It would also be helpful to improve our stopping criteria and the next step would include correlating the various methods with corresponding field data. This would be an integral part of the lifecycle analysis phase.

Taking the next step in implementing the lifecycle approach would involve extending the manufacturing—field correlation work already underway and also incorporating design qualification testing.

Finally, a long term plan might consider integrating the characterization techniques into the feedback loop of an automated adaptive process. This type of mechanism could highlight subtle changes more effectively and it may be possible to trigger corrective actions automatically.

12.   SUMMARY

The event rate modeling approach helps characterize system-level test processes and provides a visual representation of their performance. A detailed statistical evaluation allows test times to be optimized without adversely impacting outgoing quality. In many cases, apparently similar products and test processes can generate significantly different event rates. This can stimulate investigations to understand fault levels and improve fault isolation capability.

These immediate and significant manufacturing process improvements provide the motivation to adopt a consistent characterization methodology. This can then evolve to support broader lifecycle analyses with the potential to yield more substantial cost savings.

13.   ACKNOWLEDGEMENTS

The authors would like to acknowledge the advice and practical assistance of Peter Carr, Mary O'Neill, Simon Martin, John O'Toole, Paul Molloy, Mike Buckley and Richard Collins, all currently or formerly with Digital Equipment in Scotland or Ireland. They also thank Simon Martin and Jane Farren for reviewing this article and offering constructive criticism.

14.   REFERENCES

Chillarege, R. and N. S. Bowen. 1989. "Understanding Large System Failures - A Fault Injection Experiment". *IEEE, 19th International Symposium on Fault Tolerant Computing.* pp 356-363.

Chillarege, R. and R. K. Iyer. May 1987. "Measurement-Based Analysis of Error Latency". *IEEE Transactions on Computers.* Vol. C-36, No. 5.

Czeck, E. W. and D. P. Siewiorek. May 1992. "Observations on the Effects of Fault Manifestation as a Function of Workload". *IEEE, Transactions on Computers.* Vol. 41, No 5. pp559-566.

Dear, I. D., A. P. Ambler and C. Maunder. 1992. "Field Service Strategies: the use of Cost Models". In *Economics of Design and Test for Electronic Circuits and Systems.* Edited by A.P. Ambler, M. Abadir and S. Sastry, Ellis Horwood, ISBN 0-13-224767-4

Dislis, C., I.D. Dear, J. R. Miles, S.C. Lau and A.P. Ambler. 1989. "Cost Analysis of Test Method Environments". *IEEE, Proceedings of the International Test Conference.* pp875-883.

Dislis, D., March 1992. "A Financially Based Automated Advisor for Design for Test Strategy Generation". PhD Thesis, Brunel University.

Farren, D. and A. Ambler. 1995. "Cost Effective System-Level Test Strategies". *IEEE, Proceedings of the International Test Conference.* Pp 807-813.

Farren, D. and A. P. Ambler. 1994. "System Test Cost Modeling based on Event Rate Analysis". *IEEE, Proceedings of the International Test Conference.* Pp 84-92.

Goel, A. L. and K. Okumoto. August 1979. "Time-Dependent Error-Detection Rate Model for Software Reliability and other Performance Measures". *IEEE, Transactions on Reliability.* Vol. R-28, No. 3.

Gray, J. January 1986. "Why Do Computers Stop And What Can Be Done About It ?". *IEEE, Proceedings of the 5th Symposium on Reliability in Distributed Software and Database Systems.* Pp 3-12.

Krten, O. J. and D. A. Levy. 1980. "Software Modeling for Optimal Field Entry". *Proceedings of the Annual Reliability and Maintainability Symposium.* pp 410-414.

Maxwell, P. C. and R. C. Aitken. March 1993. "Test Sets and Reject Rates: All Fault Coverage's Are Not Created Equal". *IEEE, Design and Test of Computers.*

McGough, J. G., F. L. Swern and S. Bavuso. 1983. "New Results in Fault Latency Modeling". *IEEE, EASCON Conference.* pp 299-306.

Moore, T. J., 1994. "A Test Process Optimization and Cost Modeling Tool", *Proceedings of the International Test Conference.* Pp 103-110.

Ohtera, H., S. Yamada, and H. Narihisa. August 1990. "Software Availability based on Reliability Growth Models". *Transactions of the IEICE.* Vol. E 73, No. 8.

Shin, K. G. and Y. H. Lee. April 1986. "Measurement and Application of Fault Latency". *IEEE, Transactions on Computers.* Vol. C-35, No. 4.

Tegethoff, M. M. V. and T.W Chen, 1994 "Manufacturing Test Simulator: A Concurrent Engineering Tool for Boards and MCMs". *Proceedings of the International Test Conference.* Pp 903-910.

Turino, J. October 1994. "Lifetime Cost Implications of Chip-Level DFT", *Test Synthesis Seminar, IEEE International Test Conference*, TS Paper 2.2.

Wohl, J. G. May/June 1982. "Maintainability Prediction Revisited: Diagnostic Behavior, System Complexity, and Repair Time". *IEEE, Transactions on Systems, Man, and Cybernetics.* Vol. SMC-12, No 3. Pp 241-250.

Chapter 8

A Standard for Test and Diagnosis

Jack Taylor
APSYS, Ltd.

Keywords: AI-ESTATE, EXPRESS, fault tree, enhanced diagnostic inference model, system test, testability, EDIM.

Abstract: No standard exists, currently, addressing the use of AI systems in test environments. AI-ESTATE is intended to fill this void. This chapter provides an update on the status of all of the AI-ESTATE standards and their potential use to support diagnostic tools and applications. It includes a description of IEEE AI-ESTATE standard for exchanging diagnostic information and embedding diagnostic reasoners in any test environment. Also described are the defined formats and services, an example application and current industry acceptance.

1. INTRODUCTION

Recent initiatives by the Institute of Electrical and Electronics Engineers (IEEE) on standardizing test architectures have provided a unique opportunity to improve the development of test systems. The IEEE P1232 "Artificial Intelligence Exchange and Service Tie to All Test Environments (AI-ESTATE)" initiative is ushering in the next generation in product diagnostics by standardizing diagnostic services and development tool interfaces. This is being achieved by defining, developing and specifying the following:

- Problem encapsulation
- Definition of interface boundaries
- Development of standard exchange formats
- Specification of standard services applicable to the exchange formats

AI-ESTATE is intended to provide a methodology for developing diagnostic systems that will be interoperable, have transportable software, and move beyond vendor and product specific solutions. The concepts in the AI-ESTATE standard are not limited to the arena of automatic test equipment, but apply equally to manual, automatic, and semi-automatic test, as well as the domains of electronic, mechanical and pneumatic systems. The AI-ESTATE standard has been designed to abstract specific test and product details out of the diagnostic models and ties these models to domain-specific models as needed to complete the test system. This chapter includes a description of the status of the standard at the time of writing. It describes the AI-ESTATE architecture (IEEE, 1995) and the two "component" standards of the AI-ESTATE family. Within this description there is coverage of the following areas:

- A description of a neutral exchange format for diagnostic models standardized in IEEE Std 1232.1-1997 (IEEE, 1997),
- Software services provided by an AI-ESTATE conformant diagnostic system as standardized in IEEE Std P1232.2 (IEEE, 1998),
- Industry involvement and acceptance of the standards,
- Example application scenarios using these standards,
- Description and suggestions for alternative uses of the standards in various test environments,
- Discussion on the proposed new information model for dynamic data,
- Discussion and provision of directions for future work on the standard.

2.   WHAT IS AI-ESTATE?

AI-ESTATE is a set of standards under development by a subcommittee that first saw the light of day within the IEEE Standards Coordinating Committee 20, (SCC20). Its formation was during the IEEE SCC20 meeting held in Edinburgh, Scotland during May 1989. It has since grown from this beginning and is now one of the major subcommittees within the SCC20. It is intended that these standards specifically address the use of AI systems in test environments.

Recently, with the approval of IEEE Std 1232-1995 and IEEE Std 1232.1-1997, the International Electrotechnical Commission's Technical Committee 93 (IEC/TC93) accepted a proposal to advance the AI-ESTATE standard for fast track standardization at the IEC level. Current membership of IEC/TC93 includes representatives from Denmark, Finland, France, Germany, Japan, the United Kingdom, and the United States. Sponsorship for AI-ESTATE fast track came from the French delegation. The IEC/TC93 is best known for standardizing EDIF 3 0 0 and VHDL at the international level.

3. BACKGROUND TO THE DEVELOPMENT OF AI-ESTATE

The increasing complexity and cost of current systems, the inability to consistently diagnose and isolate faults in the system using conventional means and the advances in artificial intelligence technology have fostered the growth of AI technology in test and diagnosis. The proliferation of diagnostic reasoners and tools necessitates establishing standard interfaces to these tools and formal data specifications to capture relevant diagnostic information. Current test standards (e.g., Boundary Scan and STIL) (IEEE, 1990; IEEE, 1996) provide no guidance for using AI technology in test applications. Proposed AI standards (e.g., KIF) do not specifically address the concerns of the test community (IEEE, 1993). Thus, no current standard exists which addresses the use of AI systems in test environments. The AI-ESTATE standards are intended to fill this void.

The AI-ESTATE subcommittee has established several goals for the AI-ESTATE standards that include:

- Provision of a standard interface between diagnostic reasoners and other functional elements that reside within an AI-ESTATE system.
- Provision of formal data specifications to support the exchange of information relevant to the techniques commonly used in system test and diagnosis.
- Maximization of the compatibility of diagnostic reasoning system implementations.
- Accommodating of embedded, coupled, and stand-alone diagnostic systems.
- Facilitation of portability, reuse, and sharing of diagnostic knowledge.

To achieve these goals, the AI-ESTATE subcommittee proceeded to define the architecture for a standard diagnostic system and then defined component standards for information exchange and software interfaces.

4. AN ARCHITECTURE FOR DIAGNOSIS (IEEE 1232)

The AI-ESTATE architecture presented in Figure 1 shows a conceptual view of an AI-ESTATE-conformant system. AI-ESTATE applications may use any combination of functional elements and interfunction communication as shown in the figure. The service specification (P1232.2), or other specifications relevant to the particular functional element, define the form and method of communication between reasoning systems and other functional elements. AI-ESTATE identifies reasoning services provided by a diagnostic reasoner so that transactions between test system components and the reasoner are portable. AI-ESTATE assumes a client-server or cooperative processing model in defining the diagnostic services.

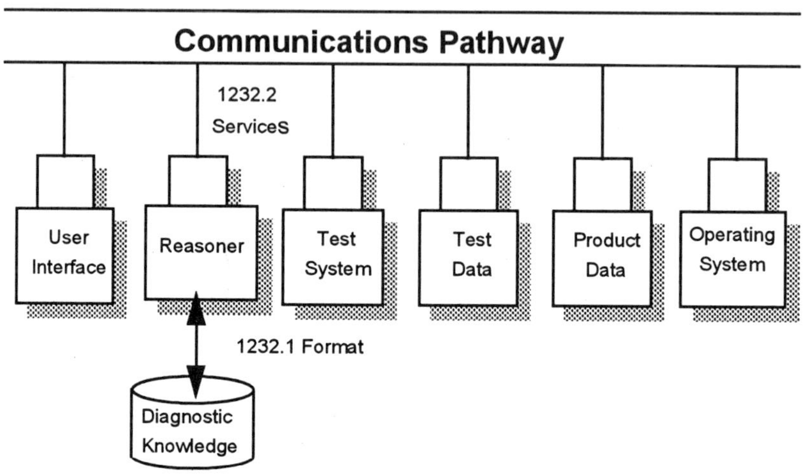

Figure 1. AI-ESTATE Architectural Concept

As indicated in Figure 1, AI-ESTATE includes two component standards focusing on two distinct aspects of the stated objectives. The first aspect concerns the need to exchange data and knowledge between conformant diagnostic systems. By providing a standard representation of test and diagnostic data and knowledge and standard interfaces between reasoners and other elements of a test environment, test, production, operation, and support costs can be reduced.

Two approaches can be taken to address this need: providing interchangeable files (P1232.1) and providing services for retrieving the required data or knowledge through a set of standard accessor services (P1232.2). AI-ESTATE is structured such that either approach can be used (IEEE, 1997; IEEE, 1998).

The second aspect concerns the need for functional elements of an AI-ESTATE conformant system to interact and interoperate. The AI-ESTATE architectural concept provides for the functional elements to communicate with one another via a "communications pathway." Essentially, this pathway is an abstraction of the services provided by the functional elements to one another. Thus, implementing services of a reasoner for a test system to use, results in a communication pathway being established between the reasoner and the test system.

AI-ESTATE services (P1232.2) are provided by reasoners to the other functional elements fitting within the architecture illustrated by Figure 1. These reasoners may include (but are not necessarily limited to) diagnostic systems, test sequencers, maintenance data feedback analyzers, intelligent user interfaces, and intelligent test programs. The current focus of the standards is on diagnostic reasoners. In addition to providing services to the test system, the human presentation system, a maintenance data collection system, and possibly an intelligent unit under test, the reasoner also uses services provided by these other systems as required. These services are not specified by the standards.

5.   MODELS FOR DIAGNOSIS (IEEE 1232.1)

The current version of IEEE Std 1232.1 defines three models for use in diagnostic systems—a common element model, a fault tree model, and an enhanced diagnostic inference model. All of the models were defined using ISO 10303-11, EXPRESS (ISO, 1994a). EXPRESS is a language for defining information models and has received widespread acceptance in the international standards communities of ISO and IEC. For example, EDIF 3 0 0 and EDIF 4 0 0 were defined using EXPRESS.

The common element model defines information entities, such as a test, a diagnosis, an anomaly, and a resource, which will be expected to be needed by any diagnostic system. The common element model also includes a formal specification of costs to be considered in the test process. A graphical view of the common element model is shown in Figure 2.

The cost model associated with the common element model is shown in Figure 3. These models are shown graphically in EXPRESS-G (ISO, 1994a). Entity relationships are shown with lines terminated by circles. For example, from Figure 2, it can be seen that diagnostic_model is composed of sets of model_anomaly, model_test, model_resource, and model_diagnosis. Dashed lines indicate the associated relationships are optional and dashed boxes indicate defined types. Heavy lines indicate a supertype relationship. For example, cost is a supertype of one of time_cost or non_time_cost. The "(ABS)" notation indicates cost is an abstract supertype, meaning the cost entity cannot be instantiated without one of its subtypes.

The remaining two models represent knowledge that may be used by specific types of diagnostic systems. The fault tree model defines a decision tree based on outcomes from tests performed by the test system (Simpson and Sheppard, 1993). Each node of the tree corresponds to a test with some set of outcomes. The outcomes of the tests are branches extending from the test node to other tests or to diagnostic conclusions (such as *No Fault*). Typically, test programs are designed around static fault trees; therefore, the AI-ESTATE subcommittee decided to include a representation for a fault tree in the standard, even though fault trees are not typically considered to be AI systems. The EXPRESS-G representation of the fault tree model is shown in Figure 4.

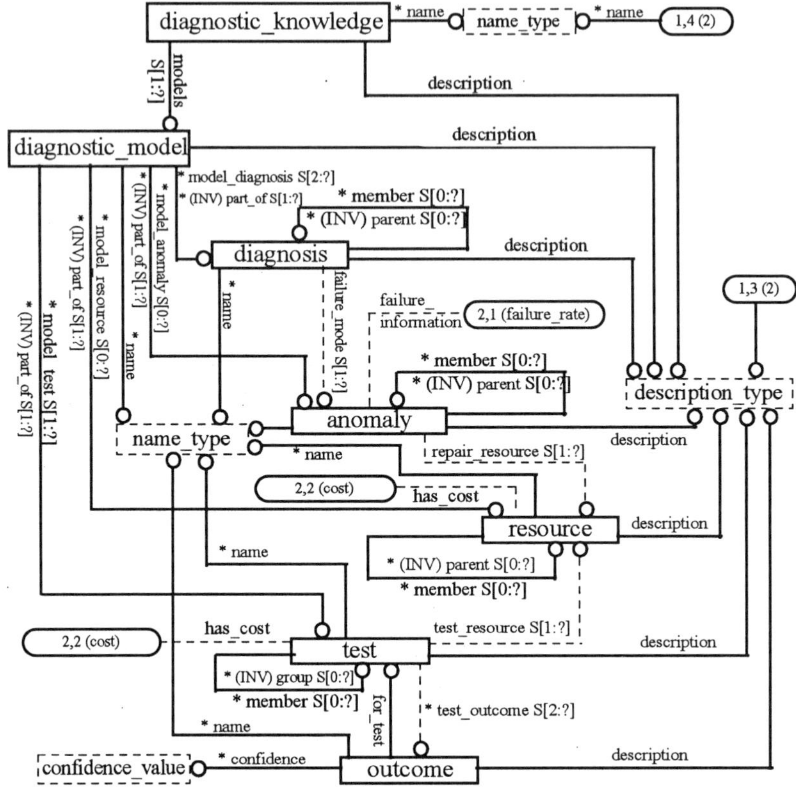

Figure 2. Common Element Model

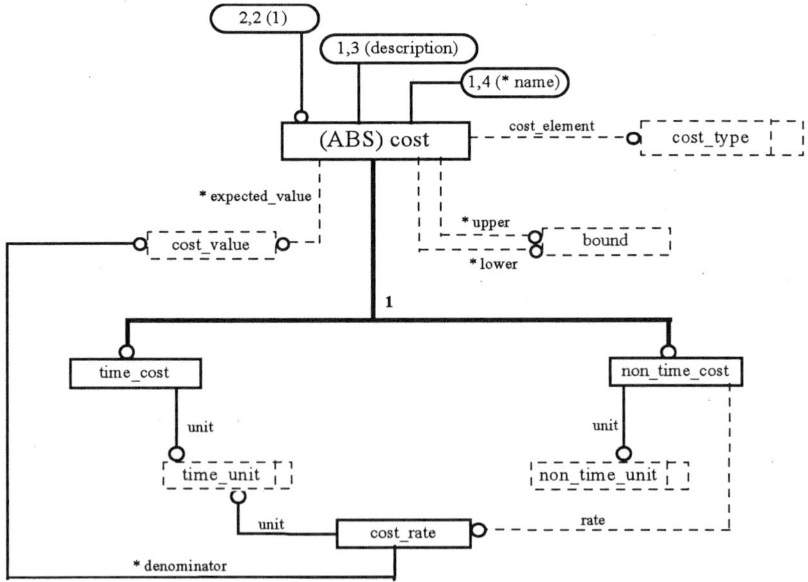

Figure 3. Cost-Model

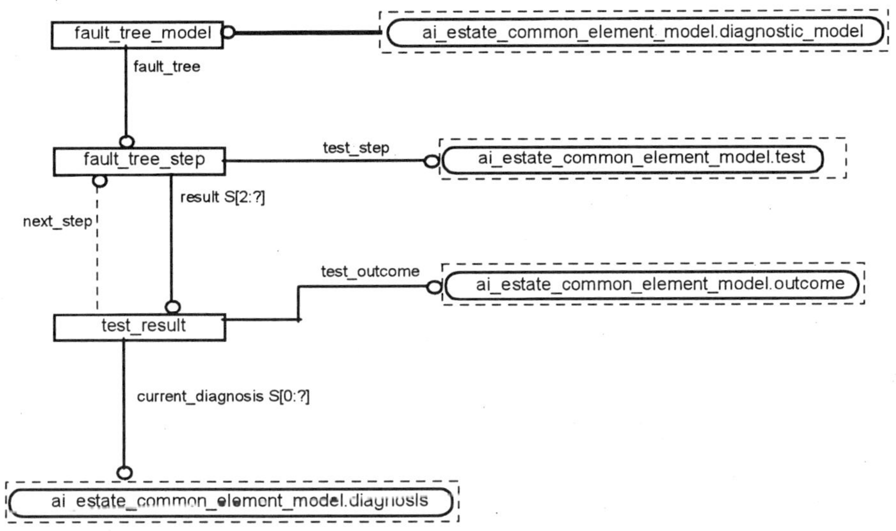

Figure 4. Fault Tree Model

The AI-ESTATE fault tree model imports elements and attributes from the common element model. Typically, test systems process fault trees by starting at the first test step, performing the indicated test, and traversing the branch corresponding to the test's outcome. The test program follows this procedure recursively until it reaches a leaf in the tree, indicating it can make a diagnosis.

The enhanced diagnostic inference model (EDIM) is based on the dependency model. Historically, test engineers used the dependency model to map relationships between functional entities in a system under test and tests that determine whether these functions are being performed correctly (Simpson and Balaban, 1982). In the past, the model characterized the connectivity of the system under test from a functional perspective using observation points (or test points) as the junctions which join the functional entities together. If a portion of the system fed a test point, then the model assumed that the test associated with that test point *depended* on the function defined by that part of the system. This type of model is also referred to as a causal model (Peng and Reggia, 1990; Pearl, 1988).

Recently, researchers and practitioners of diagnostic modeling found that the functional dependency approach to modeling was problematic and could lead to inaccurate models. Believing the algorithms processing the models were correct, researchers began to identify the problems with the modeling approach and to determine how to capitalize on the power of the algorithms without inventing a new approach to model-based diagnosis. They found that the focus of the model should be on the tests and the faults those tests detect rather than on functions of the system (Simpson and Sheppard, 1994). In particular, the focus of the model shifted to the inferences that may be drawn from real tests and their outcomes, resulting in a new kind of model called the "diagnostic inference model."

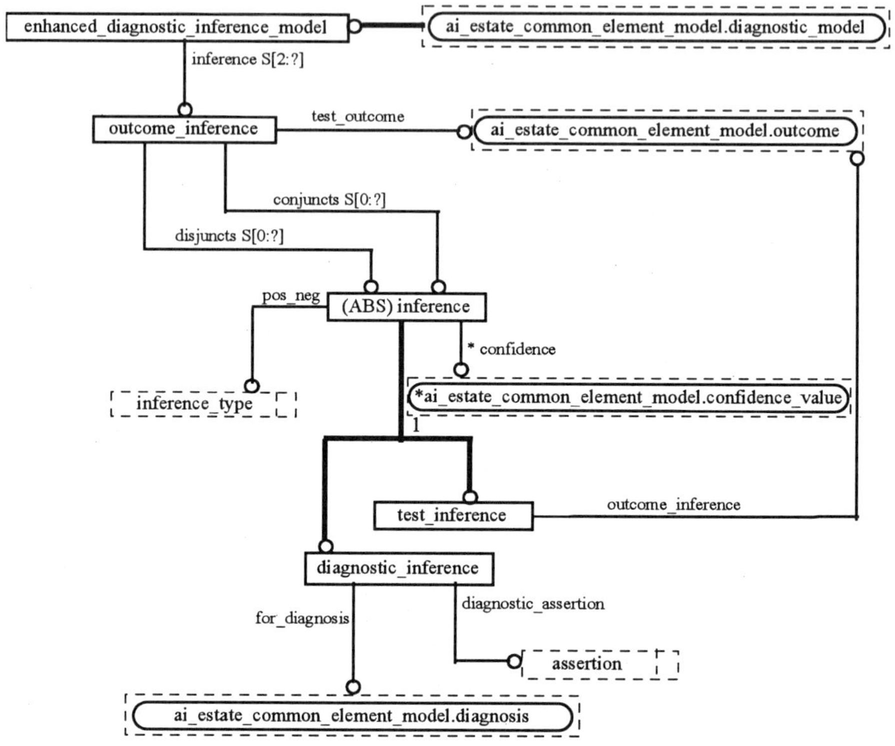

Figure 5. Enhanced Diagnostic Inference Model

The *enhanced* diagnostic inference model, defined by AI-ESTATE, generalizes the diagnostic inference model by capturing hierarchical relationships and general logical relationships between tests and diagnoses. The EXPRESS-G of the resulting model is shown in Figure 5.

The information models defined in the AI-ESTATE standard, by themselves, provide a common way of talking about the information used in diagnosis, but this is not enough for a standard. In AI-ESTATE, these models also provide the basis for a neutral exchange format. Using this neutral format, multiple vendors can produce diagnostic models in the format to enable their use by other tools that understand that format.

To specify the neutral exchange format, the decision was taken to use an instance language defined by the ISO STEP (Standards for the Exchange of Product data) community based on EXPRESS—EXPRESS-I. EXPRESS-I is an instance language defined to facilitate developing example instances of information models and to facilitate developing test cases for these models.

As an alternative, the ISO STEP community has defined a standard physical file format derived from EXPRESS models (ISO, 1994b).

Unfortunately, the STEP physical file format is very difficult for a human to read but very easy for a computer to process. The AI-ESTATE subcommittee found added benefit in EXPRESS-I over the STEP physical file format in that the language is both computer-processable and human-readable. For example, the following defines the inferences that can be drawn from a particular test of a half-adder that has passed:

```
t4_pass_implies = outcome_inference{
        test_outcome -> @t4_pass;
        conjuncts -> (@x_sa0_absent,
              @y_sa1_absent,
              @y_sa0_absent,
              @S_sa1_absent,
              @C_sa0_absent);
        disjuncts -> ();
   };
```

6. SERVICES FOR DIAGNOSIS (IEEE 1232.2)

In addition to defining models for knowledge exchange, the AI-ESTATE standard defines several software services to be provided by a diagnostic reasoner. The nature of these services enables the reasoner to be embedded in a larger test system; however, it is possible that the diagnostic system is a stand-alone application connected to a graphical user interface of some kind.

Currently, the services defined by AI-ESTATE are classified as either static model accessor services, reasoner state accessor services, knowledge acquisition services, and reasoner control services. Following the object-oriented programming paradigm, it was found that all services could be represented in one of four forms: create, get, put, or delete (Booch, 1994). Since knowledge acquisition services provide the create, put, and delete services to match the model accessor services, these will be considered together.

6.1. Static Model Traversal Services

With the publication of the data and knowledge specification (IEEE, 1997), the first set of services defined for the service specification (IEEE, 1998) focused on the existing models. The data and knowledge specification defines three models—a common element model, a fault tree model, and an enhanced diagnostic inference model (EDIM). The common element model provides definitions of basic entities expected to be used by any diagnostic reasoner, the fault tree model and EDIM organize these entities in a way to facilitate diagnostic reasoning.

The model traversal services provide the means for a reasoner to process the model on line. As an example, IEEE Std 1232.1-1997 defines, using EXPRESS (ISO, 1994a), a diagnostic model to be the following entity:

```
ENTITY diagnostic_model;
    name              :   name_type;
    description       :   description_type;
    model_test        :   SET [1:?] OF test;
    model_diagnosis   :   SET [2:?] OF diagnosis;
    model_resource    :   SET [0:?] OF resource;
    model_anomaly     :   SET [0:?] OF anomaly;
END_ENTITY;
```

Based on this model, the services can be defined for traversing the model to obtain information about a particular diagnostic model. For example, consider the set of tests defined in a particular model. To find the set of tests, there must first be access to the model itself. To obtain this access, either the name or the internal identifier of the model must be known.

The service, get_diagnostic_model(name), returns an identifier for a diagnostic model identified by its name. This identifier is then used to access the set of tests defined in the model with the service get_model_tests(model_id). This service would return a list of tests which can then be used with another service to reference the actual test entities.

Associated with each entity in the model will be the four types of services given above (i.e., create, get, put, and delete). EXPRESS includes the ability to define functions and procedures (similar to a programming language), but the intent is for these functions and procedures to be used in defining constraints on model entities (Schenk and Wilson, 1994). The AI-ESTATE committee observed, however, that EXPRESS provides a natural fit between information modeling and service definition since the type system is provided by the models. Consequently, the committee decided to use EXPRESS to define the services as well. For example, get_model_tests is defined in P1232.2 as follows:

```
FUNCTION
   get_model_tests(model_id:diagnostic_model):
      SET [1:?] OF test;
END_FUNCTION;
```

Once all of the services have been defined in EXPRESS, language bindings can be provided to make the services enable the services to be implemented by a reasoner. Currently, AI-ESTATE is exploring the development of language binding for Ada and C.

6.2 Reasoner State Services

Given a model has been loaded and a system is being tested, the model needs to be processed for diagnosis to be performed. Simply traversing the models defined in P1232.1 is not useful since the model only defines expectations rather than reality. Any reasoner will maintain an internal state of its reasoning process that will be used to report a diagnosis or to explain its reasoning. For diagnostic reasoners to be "plug-and-play" compatible, certain diagnostic state information needs to be in common between the reasoners.

To facilitate defining services for accessing or modifying the state of the reasoner, the AI-ESTATE committee made some assumptions about common diagnostic information. These assumptions were derived from the common element model and extended to support various types of reasoners.

Specifically, the AI-ESTATE committee has defined a new information model, called the *dynamic context model,* that provides the type structure for reasoner state (Figure 6).

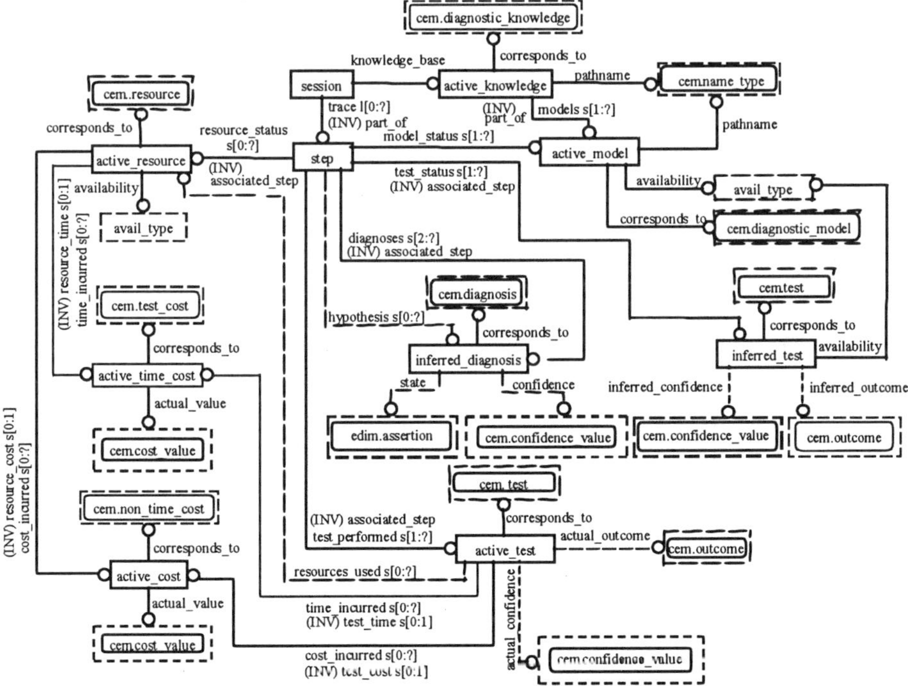

Figure 6 Dynamic Context Model

The "starting point" for the model is given by a diagnostic session. The session corresponds to a sequence of steps in which the state of the reasoner changes at each step. At each step, the reasoner keeps track of what it knows about the state of models (i.e., whether or not a model is available for processing), the resources, the tests, and the diagnoses. In addition, for each resource and test, the actual cost incurred at that step (both time and non-time) is recorded.

This model can be traversed similarly to the models defined in P1232.1. More interesting is the fact that entities can be created "on the fly" and the states of these entities vary from step to step. Another interesting byproduct of this model is that, by organizing the state according to steps in a session, the information captured can be fed to a maintenance data collection system and retained as a diagnostic history.

Another issue related to reasoner state is how one "explains" the reasoning process. During committee discussion, this was a highly contentious issue since no one could agree on a good method for explanation. The debate was settled when it was observed that the process of traversing reasoner state provides a rudimentary approach to explanation. Then, if tool developers want to provide "higher order" services, they can use the standard services to collect the necessary information. For example, explaining the currently inferred value for a diagnosis can be accomplished by the following two stages:

- Identifying the tests in the model that affect the diagnosis
- Showing the step in the session where one of these tests was performed that led to the current assertion.

6.3   Reasoner Control Services

AI-ESTATE anticipates defining several services for controlling the reasoner. Currently, five services have been identified:

1. attach_model
2. detach_model
3. save_model
4. load_model
5. select_test.

All of these services focus on either making models available, updating the models, or selecting tests for the test system to perform.

In an earlier version of the standard, several other services were provided to control the reasoner, such as setting search criteria, applying test outcomes and reverting the state to a previous state. All of these services have been deleted since they can all be handled by "putting" values in the dynamic context model. For example, applying a test outcome involves a call to put_actual_outcome. This would inform the reasoner that a new outcome is available for processing, and the reasoner state would be updated accordingly.

7.   AI-ESTATE APPLICATIONS

Since the purpose of testing is to gain information about the system under test, and diagnosis is the process of interpreting the test information, the assertion is made that all testing is done in the context of diagnosis. More simply, the only purpose a test serves is to provide an outcome that can be used to infer something about the system being tested. This premise broadens the view of diagnosis beyond the process for only isolating faults. Rather, diagnosis is considered to be the process of determining the state of the system under test relative to some anticipated state. Therefore, it is envisioned that AI-ESTATE will be applied in a variety of test contexts.

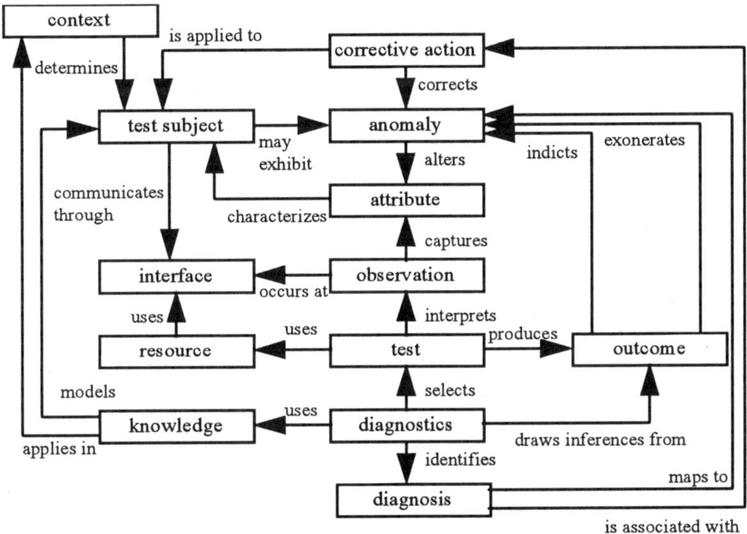

Figure 7. Model of System Test.

To capture this expanded role of a diagnostic system in test, a model of the test process has been developed. This model is described in detail in (Sheppard and Simpson, 1996a) and is reproduced here in Figure 7. In interpreting this model, note that none of the standard object (Booch, 1994; Schlaer and Mellor, 1992), activity (FIPS, 1993), or information modeling (Schenk and Wilson, 1994) techniques commonly in use today has been applied.

To read this model, each of the blocks represents entities or objects within the test environment. The relationships between the entities provide "processes" or "constraints" that one entity performs or imposes on another.

In some cases, the processes are highly dynamic (e.g., diagnostics selects test), and in others they simply define a relationship (e.g., knowledge applies_in context). In all cases, the entity to entity relationships can be read as a simple sentence (subject verb object). Given this approach the model can be used as a narrative description of the test process.

The element of this model that is of the greatest interest to AI-ESTATE is the diagnostics. As previously noted, the purpose of testing is information discovery and drawing conclusions about the test_subject. The process of drawing conclusions from test information is referred to as "diagnostics." Thus consider the diagnostic process centering on the entity labeled diagnostics. Here diagnostics uses knowledge to draw_inferences_from outcome to then identify diagnosis. As one might expect, diagnostics selects test which in turn produces outcome. Then outcome either indicts or exonerates anomaly. Once the diagnosis has been determined, one can take appropriate action since diagnosis maps_to anomaly and is_associated_with corrective_action. The whole process of performing a diagnosis is fully dependent on what diagnostics knows about the test_subject. This is captured by knowledge that models test_subject and applies_in context.

Testing can be performed under a wide variety of contexts. Indeed, the context completely defines the requirements and objectives of the test process. For example, in the context of specification compliance, testing can determine whether or not a design or a manufactured unit conforms to specifications. In the context of acceptance testing or product delivery, testing can determine if the unit satisfies a set of customer requirements. During performance or operational evaluation, testing can be used to determine optimal performance parameters for the system. In maintenance, testing can assist in detecting and isolating faults.

The primary advantage of providing an abstract model for test and diagnosis and associating that model with a "system view" of the test subject is that it provides a means for abstracting details out of the test problem until needed. This in turn permits optimizing the structure of the test problem, thus managing the complexities inherent in manipulating large, heterogeneous systems. For this to work, it is essential to be able to see how to map real systems into the model. This in turn enables us to identify the critical elements of the test problem and to optimize the solution. In this section, the mapping of two "systems" into the model is used to illustrate.

The first system considered is very simple. Consider an integrated circuit containing four D flip-flops (e.g., an SN5475). The test_subject would correspond to this IC, it is then required to be able to identify the attributes of the IC as the nominal behavior of the chip. This would usually be found in the truth table for each of the flip-flops. In addition, the adding of characteristics such as voltage, temperature range, pin orientation, size and color of the packaging, etc. The actual set of attributes of interest would depend on the context under which testing is taking place. If the point of interest is failure modes associated with the pins and the logic of the chip, then the focus of interest will be to the logic values observed at the pins. If interest is in manufacturing issues then characteristics of the packaging may be included.

For the sake of discussion, the context is limited to assessing the logical performance of the chip at the pins. Thus, our attention will be restricted to the logic specification and to the set of stuck-at faults. This set of faults would define the anomalies that might be exhibited by the chip. If a digital tester with a fault dictionary is used, then knowledge would correspond to the fault dictionary itself and diagnostics would include the test controller and the matching algorithm in the dictionary. The tests would be defined by the vectors in the fault dictionary (actually, each of the output values for each vector would correspond to a different test). Whenever a vector is processed, the associated tests would either pass or fail (thus the outcomes are known) and these outcomes would point to the possible anomalies of the chip. In the event the chip is faulty, possible corrective_actions include replacing the chip or re-setting (or re-soldering) the chip in the socket.

The role AI-ESTATE plays in this environment is two-fold. First, a model conforming to the EDIM defined by 1232.1 can be used to capture the diagnostic knowledge resident in the fault dictionary (Sheppard and Simpson, 1996b). This means that the diagnostic knowledge inherent in fault dictionaries can be exchanged in a standard way between test systems by using IEEE Std 1232.1. When coupled with another standard such as P1029.1 (WAVES), P1445 (DTIF), or P1450 (STIL), complete digital test information can be tied to an encapsulated diagnostic system to facilitate efficient and accurate diagnostics.

The second role for AI-ESTATE facilitates a means for replacing the traditional, but often proprietary, fault dictionary signature matching process. By providing standard services for applying and interpreting test results, an AI-ESTATE-conformant reasoner can replace the signature matcher.

For the second system example, consider a case at the other end of the spectrum. Consider a network of satellites in orbit that provide the backbone for a communications network. One test context of interest would be to verify that all of the defined links in the network are present (including satellite-to-satellite cross links and ground links). Thus, the definition of our test_subject would be the connectivity of the network. Attributes of this system might include transmission of messages between two points in the network in a reasonable period of time and then define tests to be a set of messages that traverse the network in some predictable way. The anomalies would correspond to links being down.

For this system, the knowledge required for diagnosis and the associated diagnostics might be significantly different from the fault dictionary of the first system. For example, it is possible to rely on a set of

SNMP (simple network management protocol) traps to signify when a link drops in the network (assuming the endpoints of the links are treated as SNMP managed objects). In this case, the knowledge would correspond to the MIB (managed-object information base), and the diagnostics would be a passive network monitoring system or alarm correlator.

Here, the AI-ESTATE application is also two-fold. Many network systems and associated network management systems utilize heterogeneous software technology. To date, they have focused on information provided by standard protocols and specifications of managed objects and not on capturing information relating network alarms to possible causes. Similar to the fault dictionary discussed above, models to be used in alarm correlation can be developed, merged, and exchanged between applications in a network environment using IEEE Std 1232.1. Similarly, the alarm correlation and fault isolation services that will be required in network fault management and trouble ticketing systems can utilize IEEE Std 1232.2.

8.   INDUSTRY INVOLVEMENT AND ACCEPTANCE

The AI-ESTATE standard has been under development for six years under the sponsorship of three IEEE societies—the Computer Society, the Aerospace Electronic Systems Society, and the Instrumentation and Measurement Society. Current membership on the subcommittee includes representatives from academia (University of Connecticut and Johns Hopkins University), government (U.S. Navy, U.S. Air Force, U.S. Army), and industry (Boeing, NCR, AlliedSignal, IET Intelligent Electronics, ARINC, Etec). In addition, the subcommittee has representatives from several countries other than the United States (United Kingdom, France and Germany).

In addition to the broad support within the standards community, several tool vendors are committed to enhancing or providing tools that conform to the AI-ESTATE standard. Several tools already implement pre-standard versions of the AI-ESTATE models including:

- ARINC's System Testability and Maintenance Program (STAMP)
- ARINC's Portable Interactive Troubleshooter (POINTER)
- Detex's System Testability Analysis Tool (STAT)
- IET's TechMate
- Giordano Associates' Diagnostician
- Qualitech's Testability Engineering and Maintenance System (TEAMS)
- The Navy's Integrated Diagnostic Support System (IDSS).

These tools are likely to be upgraded to the AI-ESTATE standards in the near future.

## 9.	CURRENT STATUS OF AI-ESTATE

The status (as of June 1998) of the AI-ESTATE series of standards is tabulated below:

- IEEE Std 1232-1995. 1995. Standard for Artificial Intelligence Exchange and Service Tie to All Test Environments (AI-ESTATE): Overview and Architecture.
- IEEE Std 1232.1-1997. 1997. Trial Use Standard for Artificial Intelligence Exchange and Service Tie to All Test Environments (AI-ESTATE): Data and Knowledge Specification.
- IEEE P1232.2. 1998. Trial Use Standard for Artificial Intelligence Exchange and Service Tie to All Test Environments (AI-ESTATE): Service Specification, Draft 4.3.
- IEEE P1522. Trial Use Standard for Testability and Diagnosability Characteristics and Metrics, Work commenced on this topic in November 1997 and the first draft is expected by November 1998.

## 10.	FUTURE DIRECTIONS FOR AI-ESTATE

There is currently work progressing on a number of topics. These are listed below under their various headings. The current state of these topics is addressed and explanations given to the ideas and reasons behind them.

### 10.1	Testability Standard

As defined in MIL-STD-2165, testability is "a *design characteristic* which allows the status (operable, inoperable, or degraded) of an item to be determined and the isolation of faults within the item to be performed in a timely manner." The purpose of MIL-STD-2165 was to provide uniform procedures and methods to control planning, implementation, and verification of testability during the system acquisition process by the Department of Defense (DoD). It was to be applied during all phases of system development—from concept to production to fielding. This standard, though deficient in some areas, provided useful guidance to government suppliers. Further, lacking any equivalent industry standard, many commercial system developers have used it to guide their activities though it was not imposed as a requirement.

There is a need for a new industry standard that addresses system testability issues and that can be used equally well by both commercial and government sectors. This is required because of the following reasons:

- The current emphasis within the DoD on the use of industry standards.
- The continuing need to control the achievable testability of delivered systems in DoD and commercial sectors.
- The removal of MIL-STD-2165 as a standard with no replacement (commercial or DoD).

To be useful, this commercial standard must provide specific, unambiguous definitions of criteria for assessing system testability. In addition, the standard should recommend processes for implementing a testability program as part of a product's full life cycle.

MIL-STD-2165 was deficient in the precise definition of measurable testability figures-of-merit and relied mostly on a weighting scheme for testability assessment. (It should be noted, however, that the standard did permit the use of analytical tools for testability assessment such as SCOAP, STAMP, and WSTA.) Now that a standard diagnostic model exists (IEEE, 1997) and the definition of reasoner services are nearing completion (IEEE, 1998), AI-ESTATE has decided to explore creating a standard defining testability metrics in terms of the knowledge and service standards.

In the industry, many terms such as test coverage and fault detection are not well defined and not comparable from system to system. Some measures, such as false alarm rate, are not measurable in field applications. An immediate benefit will come with a consistent, precise, measurable set of testability attributes that can be compared across systems and within iterations of system design.

The ability of a system implementation to support its fault isolation goals must be measured by:

- The fundamental capability of the diagnostic model to support the reasoning process.
- The ability of the reasoner to do the reasoning.

For example, suppose that a diagnostic system is intended to work in a time critical environment. Furthermore, suppose the diagnostic model contains the information required to fully fault isolate to the required level unambiguously. If the reasoner is too slow, then the system will not satisfy its performance requirements. Similarly, if the diagnostic model or the set of tests supporting that model contains insufficient information to achieve fault isolation to the required level, it will not matter how fast the reasoner is. The system will be incapable of satisfying its isolation requirements. Metrics must be established that allow us to clearly identify important testability parameters of both models and reasoners.

MIL-STD-2165 attempted to standardize both programmatic and measurement tasks. Variations in the internal program management methods of different companies and DoD organizations make the standardization of the programmatic tasks unreasonable. The subcommittee is exploring creating a "Recommended Practice" document to support the programmatic aspects of system testability. MIL-STD-2165 identified five essential elements of a comprehensive testability program, including the following:

1. Testability program plan
2. Establishment of achievable testability requirements
3. Participation in the design process
4. Prediction and evaluation of design testability
5. Inclusion of testability in program reviews.

Lacking well-defined testability measures, the task of establishing testability requirements then predicting and evaluating the testability of the design is extremely difficult. This in turn makes effective participation in the design for testability process difficult. These difficulties will be greatly diminished by the establishment of standard testability metrics. Beyond the participation with design engineering, MIL-STD-2165 also puts emphasis on the information exchanged between the testability and logistics functions and the testability and maintainability functions.

As we strive to establish concurrent engineering practices, the interchange between the testability function and other functions becomes even more important. To create integrated diagnostic environments, where the elements of automatic testing, manual testing, training, maintenance aids, and technical information work in concert with the testability element, maximization of the reuse of data, information, knowledge, and software will be essential. Complete diagnostic systems include BIT, ATE, and manual troubleshooting. It would be desirable to be able to predict and evaluate the testability of systems at these levels.

Currently, the AI-ESTATE subcommittee is gathering information from the DoD and industry about model representations, their associated metrics, and the processes put in place to utilize them. The results of this review will form the basis for defining the metrics to be included in the standard and the procedural guidance to be included in the "Recommended Practice."

10.2   Unification of the Current 1232 Set of Standards

IEEE 1232–Overview and Architecture has been affirmed as a Full Standard. The trial use standard for IEEE 1232.1–Data and Knowledge Specification will expire, (after a one-year extension) in the year 2000. The trial use standard IEEE 1232.2, (assuming it is published by the end of 1998) will expire also in the year 2000 unless a one-year extension in applied for and granted. In the light of the above, it is the intention of the subcommittee to work toward unifying all three components of the standard.

It is then planned to publish them all together as a single Full Use standard in the year 2000.

10.3   General

A large amount of work has been done on the AI-ESTATE standards and there is now widespread acceptance of the standards in government and industry. However, there remains more work to be done to keep the standards in pace with technology advances. Currently, several projects are underway within the AI-ESTATE subcommittee to do just that.

The development of models for rule-based systems and connectionist systems is also anticipated. Rule-based systems provided the first success stories in diagnosis and artificial intelligence. While the EDIM would be able to capture the logical information contained in a rule base, a separate model is required that is tailored to the rule-based architecture.

In addition, neural networks and fuzzy systems have shown tremendous promise in diagnostics, especially in the presence of noise and error. While it is unclear at this time what a standard representation of a neural network would look like, it is a worthwhile endeavor to explore the possibility of defining such a standard.

As well as defining additional models for knowledge exchange, there will also be the need to define more standard services for accessing and manipulating these models. Thus, the development of services and the development of data models will proceed in parallel. While objectives for these two activities can be separated, the common role the information plays in both contexts necessitates developing the models and the services together.

As discussed earlier, diagnosis occurs within some context. In fact, the context is central to determining the scope of the test and diagnosis problem. Unfortunately, capturing information about context in a standard way is problematic. The number of variables associated with context is excessive, and the relationships between those variables are frequently unknown. Nevertheless, the proper interpretation of diagnostic and test information relies upon a common understanding of the context in which testing takes place. Consequently, the subcommittee is beginning to develop a model of context for diagnosis. Recent work in non-monotonic reasoning and categorical reasoning offer promise in formalizing context for this problem (Akman and Surav, 1996).

11.   CONCLUSION

Reasoning system technology has progressed to the point where electronic systems are employing artificial intelligence as a primary component in meeting system test and verification requirements. This is giving rise to a proliferation of AI-based design, test, and diagnostic tools.

Unfortunately, the lack of standard interfaces between these reasoning systems is increasing the likelihood of significantly higher product life-cycle cost. Such costs would arise from redundant engineering efforts during design and test phases, sizable investment in special-purpose tools, and loss of system configuration control.

The AI-ESTATE standard promises to facilitate ease in production testing and long-term support of systems as well as reducing overall product life-cycle cost. This will be accomplished by facilitating portability and knowledge reuse and sharing of test and diagnostic information, among embedded, automatic, and stand-alone test systems within the broader scope of product design, manufacture, and support.

12.   ACKNOWLEDGMENTS

I would like to thank all the members of the AI-ESTATE subcommittee of SCC20 and the many reviewers of the AI-ESTATE standards. Especially, I thank Randy Simpson, Tony Bartolini, John Sheppard and Les Orlidge, without whose help and guidance this chapter would never have been written.

13.   References

Akman, V. and M. Surav. 1996. "Steps Toward Formalizing Context," *AI Magazine*, 17(3):55–72.

Booch, G. 1994. *Object-Oriented Analysis And Design With Applications*, 2nd Ed. Benjamin Cummings.

FIPS-183. 1993. *Integrated Definition for Function Modeling (IDEF0)*. National Institute of Standards and Technology.

IEEE Std 1029.1-1997. 1997. *Waveform and Vector Exchange Standard (WAVES)*, Piscataway, New Jersey: IEEE Standards Press.

IEEE Std 1149.1-1990. 1990. *Standard Test Access Port and Boundary Scan Architecture*, Piscataway, New Jersey" IEEE Standards Press.

IEEE Std 1232-1995. 1995. *Standard for Artificial Intelligence Exchange and Service Tie to All Test Equipment (AI-ESTATE): Overview and Architecture*, Piscataway, New Jersey: IEEE Standards Press.

IEEE Std 1232.1-1997. 1997. *Trial Use Standard for Artificial Intelligence Exchange and Service Tie to All Test Environments (AI-ESTATE): Data and Knowledge Specification*, Piscataway, New Jersey: IEEE Standards Press.

IEEE P1232.2. 1998. *Trial Use Standard for Artificial Intelligence Exchange and Service Tie to All Test Environments (AI-ESTATE): Service Specification*, Draft 2.2.

IEEE P1252. 1993. *Standard for a Frame Based Knowledge Representation*, Draft 2.1.

IEEE P1450. 1996. *Standard Test Interface Language (STIL)*, Draft 0.23.

ISO 10303-11:1994. 1994. *Industrial Automation Systems and Integration—Product Data Representation and Exchange—Part 11: EXPRESS Language Reference Manual*, Geneva: ISO Press.

ISO 10303-21:1994. 1994. *Industrial Automation Systems and Integration—Product Data Representation and Exchange—Part 21: Clear Text Encoding of the Exchange Structure*, Geneva: ISO Press.

Pearl, J. 1988. *Probabilistic Reasoning in Intelligent Systems: Networks of Plausible Inference*, San Mateo, California: Morgan-Kaufmann Publishers.

Peng, Y. and J. A. Reggia. 1990. *Abductive Inference Models for Diagnostic Problem-Solving*, New York: Springer-Verlag.

Schenk, D. A. and P. R. Wilson. 1994. *Information Modeling: The EXPRESS Way*, New York: Oxford University Press.

Schlaer, S., and Mellor, S. L. 1992. *Object Lifecycles: Modeling the World in States*, Englewood Cliffs, New Jersey: Yourdon Press.

Sheppard, J. W. and W. R. Simpson. 1996. "A Systems View of Test Standardization," *Proceedings of AUTOTESTCON '96*, Dayton, Ohio, pp. 384–389.

Sheppard, J. W. and W. R. Simpson. 1996. "Improving the Accuracy of Diagnostics Provided by Fault Dictionaries," *Proceedings of the 14th IEEE VLSI Test Symposium*, Los Alamitos, California: IEEE Computer Society Press, pp. 180–185.

Simpson, W. R. and H. S. Balaban. 1982. "The ARINC System Testability and Maintenance Program (STAMP)," *Proceedings of AUTOTESTCON '82*, Dayton, Ohio.

Simpson, W. R. and J. W. Sheppard. 1993. "Fault Isolation in an Integrated Diagnostic Environment," *IEEE Design and Test of Computers*, 10(1):52–66.

Simpson, W. R. and J. W. Sheppard. 1994. *System Test and Diagnosis*, Norwell, Massachusetts: Kluwer Academic Publishers.

Chapter 9

Advanced Onboard Diagnostic System for Vehicle Management

Kirby Keller, Jeffery Holland, Darrell Bartz, Kevin Swearingen
The Boeing Company

Keywords: Vehicle management, diagnostics, prognostics, health management, could not duplicates, false alarms, knowledge discovery, data mining, neural networks, fuzzy logic.

Abstract: Could Not Duplicates (CNDs) are the bane of modern aircraft. Inflight Built-In-Test (BIT) or pilot squawks often indicate faults that can not be duplicated with ground testing. For many subsystems CND rates run over 50 percent. The Vehicle Systems Technology (VST) group at the Boeing Company is developing an onboard diagnostic system architecture and supporting technologies to reduce CNDs and exploit the promise of prognostics. The approach involves the use of neural network and fuzzy logic models of subsystem health, correlation of health incidents with operational context and fusion of health reports within and across subsystems. Knowledge Discovery and data mining techniques are used to identify health/fault signatures and develop diagnostic/prognostic models from these signatures. This work was sponsored by Boeing internal research, the Navy through the On-line Flight Control Diagnostic System program (with Lockheed-Martin Control Systems as a subcontractor) and the Joint Strike Fighter technology maturation program.

1. INTRODUCTION

Could Not Duplicates (CNDs) are the bane of modern aircraft. Inflight Built-In-Test (BIT) or pilot squawks often indicate faults which can not be duplicated with ground testing. CND rates run over 50 percent for many subsystems. The Vehicle Systems Technology (VST) group at the

Boeing Company is developing an Advanced Onboard Diagnostic System (AODS) and supporting technologies with the goal of reducing CNDs by 50% and begin to exploit the promise of prognostics. This chapter provides an overview of this research. More complete documentation is available in program and internal research reports (Keller, 1997 and Bartz, 1997).

Analysis of field data shows that the majority of flight control CNDs stem from chronic bad actors. These bad actors manifest themselves through a sequence of fault reports that spawn a series of CNDs and improper fixes before eventually leading to an appropriate removal and replacement action. The task of reducing CNDs is strongly related to prognostics in that both require monitoring system health over a range of conditions, sensitivity to operational context under which "fault" conditions first begin to appear, and the use of trend analysis rather than more simplistic fail/no fail tests typical of current BIT. This leads to an overall "health management" capability in which the prognostic capability can be leveraged to institute scheduled, condition based maintenance.

Analysis of field data plus expert opinion drove this research to the design of an Advanced Onboard Diagnostic System (AODS). AODS is a collection of software modules, which implement on-line component or subsystem level diagnostics, and a system level element to integrate the results of the subsystem diagnostics. Affordability dictates that AODS software not be flight critical; it may integrate BIT results but AODS is strictly intended to support maintenance activity only. Neural network and fuzzy logic technologies are used as a means to create low cost models of subsystem/component health. Data mining and visualization technologies plus data collection capabilities built into the AODS design serve to close the loop and reduce the cost of development and maturation. See Figure 1.

This work is currently an ongoing effort sponsored by Boeing internal research. Previous funding sources included the Navy through the On-line Flight Control Diagnostic System program (with Lockheed Martin Control Systems as a subcontractor) and the Joint Strike Fighter technology maturation program. AODS concepts were demonstrated in the flight control laboratory with hardware in the loop. Additional workstation demonstrations prototyped applications to engine and navigation health monitoring using data from simulation models. A cost benefit analysis was begun to determine the payoff and costs associated with this approach and its eventual distribution between the aircraft and flightline.

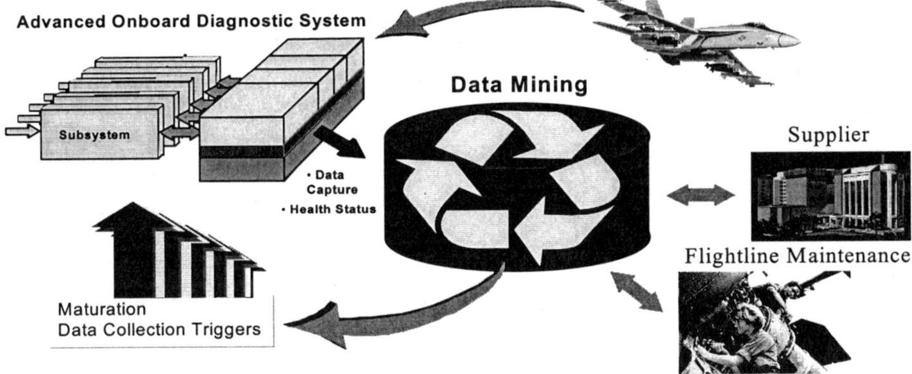

Figure 1 - AODS and Supporting Technologies

2.   AODS REQUIREMENTS

The requirements for AODS and supporting technologies were derived from an analysis of flight data and flightline maintenance actions. This analysis began with flight controls in the OFCDS program with a focus on eliminating CNDs. It was later expanded to the engine and navigation systems and to address more general health monitoring (diagnostics/prognostics) capabilities.

A number of fault classes associated with flight controls, navigation and the engine were identified to drive the analysis and later for demonstration and testing. The selection of fault classes was driven by the prevalence of CND occurrences and the opportunity to demonstrate health monitoring given the facilities and simulation models available. The following fault classes were included:

- Flight control faults (primarily actuators) which produce sequences of maintenance actions over 5-10 flights and result in a number of CNDs or improper maintenance actions. In a gross sense, the sequence of incidents detected by current BIT systems is indeed a form of prognostics that is used by experienced maintenance personnel.
- Intermittent faults typical of an open circuit under high temperature conditions within flight controls components or other electronic equipment.
- Isolated incidents that did not to reoccur.
- Faults that correlate to flight conditions such as launch or high Gs.
- Engine stalls/shutdowns and correlation to flight control health incidents.
- Navigation system health status reporting.

A combination of expert opinion, field data collected from the F/A-18 Digital Storage Unit, laboratory test benches and simulation models provided the basis for characterizing health/fault conditions.

The flight control fault class was dominated by the occurrence of a series of maintenance actions triggered by BIT and/or pilot squawks but not repeatable with ground tests. These sequences consistently lead to an eventual removal and replacement of an actuator. In almost all cases, part way through this sequence the experienced maintainer suspects an actuator is probably at fault. However, interim actions are usually taken that are less time consuming for the maintainer to perform. They include swapping or replacing the flight control computer, ignoring a BIT indication or writing it up as a CND. BIT ambiguity, lack of confidence in BIT, and pressure to keep the aircraft flying contribute to this practice. Addressing this type of fault class leads directly to a requirement for a prognostic capability. The actuator exhibits a history of degraded performance usually under stress (thermal or load) which is not repeatable with ground tests.

Fault/health monitoring of turbine engines and navigation systems demonstrated similar characteristics that led this effort to develop a general approach that would include on-line, onboard health monitoring to support prognostics and condition based maintenance. Furthermore, the AODS concept was extended to allow the distribution of the system within the aircraft and between the on-line, onboard and flightline elements in accordance with the requirements of the application. The following top level AODS capabilities and features were identified:

- Low cost software models to monitor subsystem health on-line. Budget to develop and test diagnostic software over and above flight critical BIT will be scarce in any development program. The higher resolution health monitoring software must buy its way onto the aircraft which means low development and maturation cost.
- Low cost software models to monitor operational context (e.g. launch, temperature profiles or hinge moment) and to capture context data associated with a fault/health event. Maintenance personnel uniformly complain that data on the operational conditions at the time of a fault code are either not available or not easily accessible. Collecting large amounts of data is not an acceptable solution without a means to easily use it.
- System and subsystem module partitioning, which enables software reuse, and the distribution of the system within the aircraft and between the aircraft and the flightline. To reduce cost, the health monitoring system should be able to fit easily into a range of hardware architectures or maintenance processes.

- Support for incremental development and maturation that enables a cycle of continuous improvement is a must particularly in two level maintenance environments. The AODS architecture must minimize the cost of updating health management software and provide feedback on component/subsystem performance to suppliers.
- AODS must have the capability to prioritize and focus limited processing resources on more important and urgent processes which means it should be able to make the best use of the processing resources and storage space available.
- While the focus of AODS is on higher resolution health monitoring of components, it should also include the ability to account for and leverage interactions between subsystems/components.
- The capability to fuse and perform trend analysis on health status indications over time and from multiple sources. This includes the ability to determine and weigh the confidence of health reports as a function of operational environment.
- Maintenance decision support aids that enable the maintenance crew to take advantage of the knowledge of trends in component/subsystem condition and tradeoff factors such as operational need, time to repair and safety/likelihood of mission abort.

3.   AODS ARCHITECTURE

The top level AODS architecture consists of a system module and multiple subsystem or local diagnostics modules as shown in Figure 2. The subsystem modules perform the real-time monitoring of the target subsystem and report health status. The system level module integrates health status reports and interprets the results in light of the context in which the health incident was reported. Associated with the system level diagnostics are context models that monitor operational conditions using on-line data that is not available to the subsystem models or that monitors operational context that is pertinent to several subsystem modules.

The system data flow is shown in Figure 3. The subsystem modules process real-time subsystem parameters and provide a running assessment of system health. The subsystem module reports health status in the form of an incident type, time, status indicator, frequency/duration and a confidence. Any additional data (e.g. memory inspect) that may support later processing by the system module or ground testing is also included.

The synchronization module captures appropriate system context variables or template status along with the strength of correlation to the health/fault incident. Examples of templates include aircraft launch or hinge moment. The resulting health status record is processed by the system diagnostic assessor. The diagnostic assessor is a rule based system that processes the heath status reports from the subsystem modules. It maintains a record of system health from the result of fusing previous health status messages. This health record is the basis for maintenance recommendations that can be generated on or offboard.

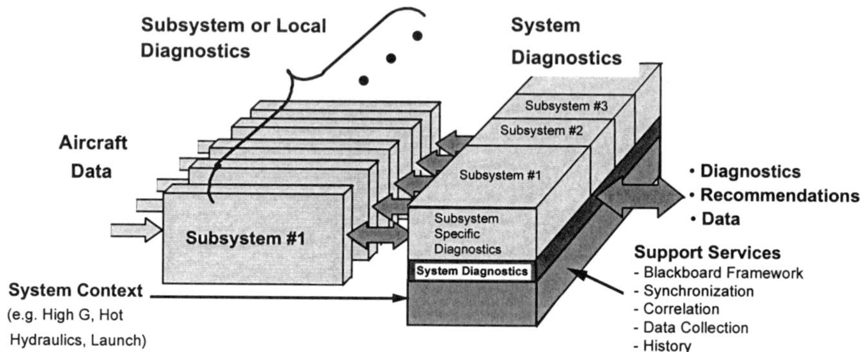

Figure 2 – AODS On-line Architecture

The reporting of a health incident is triggered by a change in health status or by the context monitors indicating a desirable context to check subsystem health. Examples of these health monitors include hinge moment limit (for actuator health) and straight and level flight (for navigation health) monitors. Changes in system health push a health report to the system level. The system level may pull a health report from the subsystem level based on desired or preferred test conditions.

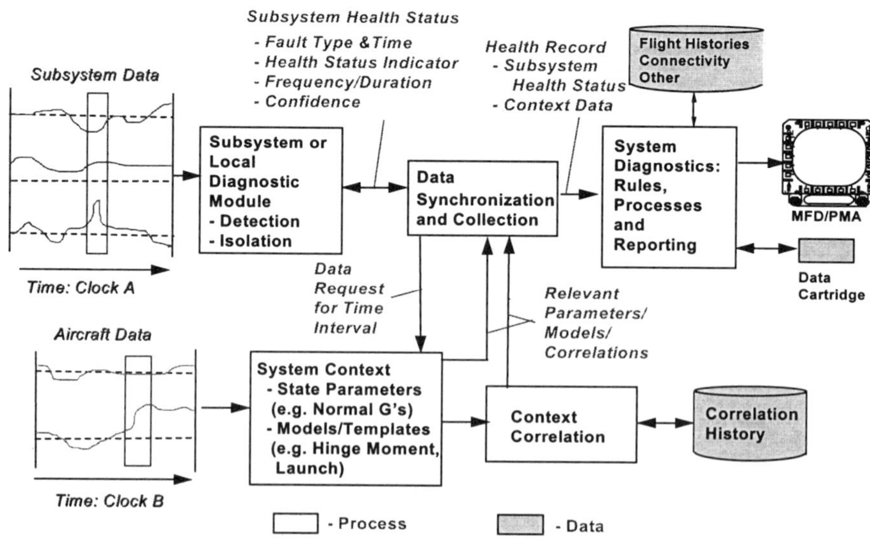

Figure 3 - AODS Top Level Data Flow

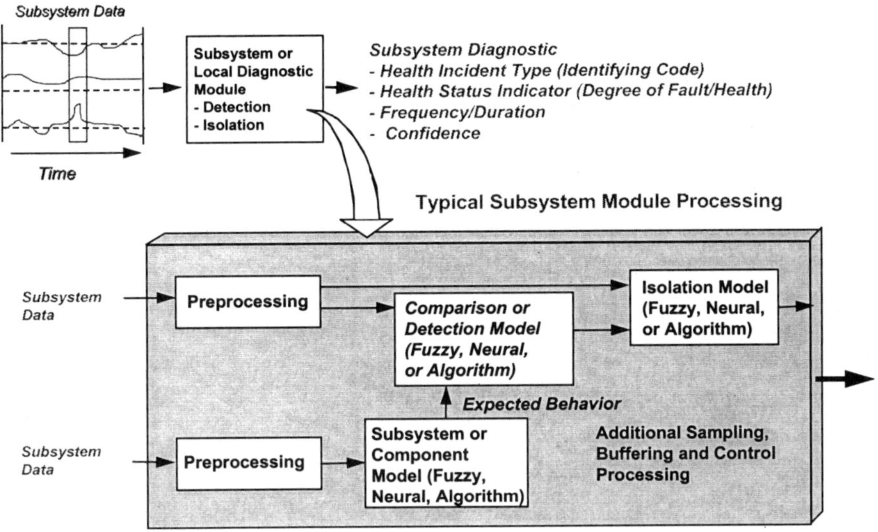

Figure 4 - Generic Subsystem Diagnostic Module

A template for the subsystem model is shown in Figure 4. Typically a neural or fuzzy model is used to generate a notion of normal or healthy subsystem performance which is compared to actual subsystem outputs. Additional models are used to identify and determine the degree of a fault/health of a particular aspect of a subsystem.

Neural networks and fuzzy systems were used as a means to model subsystem behavior. Neural networks were used for functional modeling. For example, hinge moment limit and engine stall margin estimates were represented in the form of neural networks. Neural networks were also used in the identification and estimation of system health. Fuzzy logic was used in a similar manner but developed manually rather than trained from test cases. Processes to develop fuzzy systems from data are being pursued. In comparison, the neural networks provide greater resolution than the fuzzy systems but do not lend themselves to examination and adjustments based on expert human knowledge. For this reason, the neural networks found were more applicable to functional modeling and the mapping of fault patterns to a system health indication. Fuzzy models were used in determining event correlation and in system health monitoring where the monitors were developed and adjusted based on expert knowledge.

Figure 5 shows the AODS functionality mapped into a layered architecture that includes maintenance decision aids. These decision aids support the crew in determining what maintenance actions to take and in scheduling these actions. This layer also supports continuous evolution/ maturation of the system capability, including prognostics and condition based maintenance. It is envisioned that the maintenance decision support capabilities would be embedded in upgrades to current debrief systems such as the Aviation Maintenance Environment (AME).

The health assessment layer is a key factor in the AODS concept that supports the maintenance decision process. Health incidents are handled like track files in sensor data fusion process. A single incident is generally just one input to the overall process of determining system health. Using this architecture modules or components can be distributed within the aircraft and between the aircraft and the ground support system to meet the requirements of a particular application.

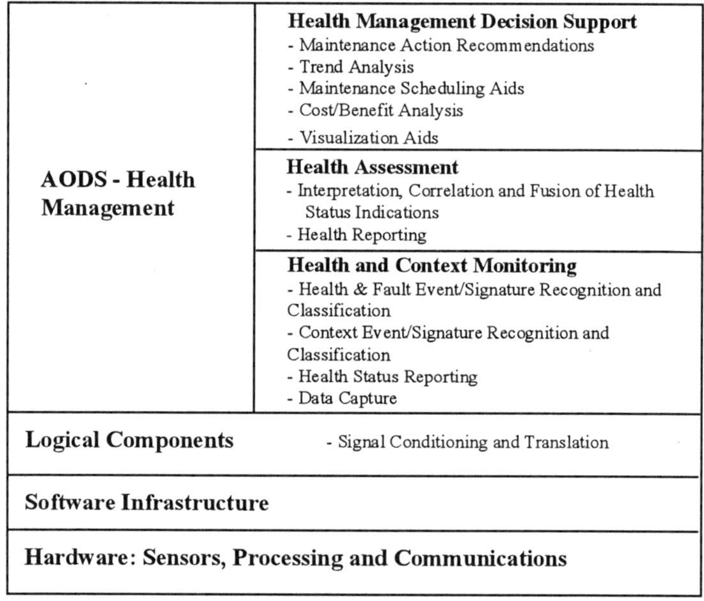

Figure 5 – AODS Health Management Architecture

4.   DEMONSTRATIONS

The AODS demonstration implemented diagnostics for flight control, navigation and propulsion subsystems and was hosted on workstations and written in Ada and C. Demonstration scenarios consisted of data streams which would be available to an onboard diagnostic system. In addition to the subsystem data, an aircraft flight data stream was generated to model the corresponding flight context or conditions for that scenario. The aircraft flight data stream was generated using a six degree of freedom simulation. Subsystem data streams were generated using the F/A-18 Integration Testbench laboratory data (See Figure 6) or from subsystem simulation models and included the following:

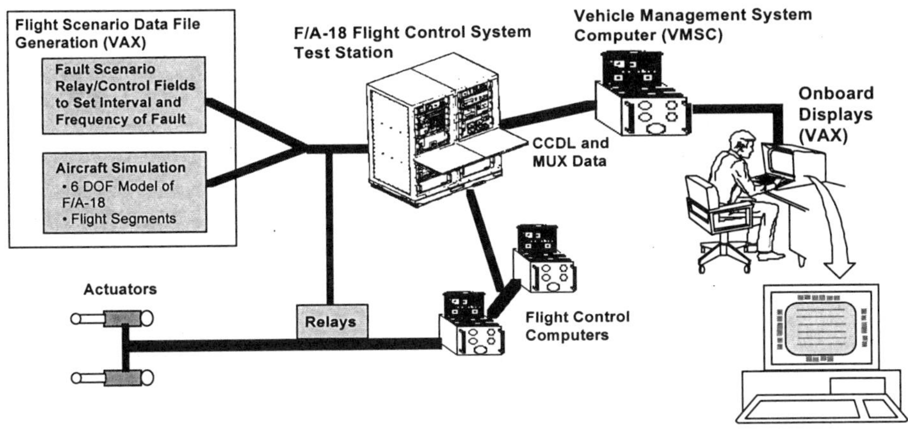

Figure 6 - F/A-18 Integration Test Bench Used in AODS Demonstration

Flight control: As part of the OFCDS program, a portion of the flight control diagnostics system was implemented in a vehicle management system computer (VMSC). The VMSC is a flightworthy processor and was integrated into the F/A-18 Integration Test Station. A 6 DOF model of the F/A-18 was used to simulate aircraft flight conditions. Fault scripts controlled the interval and frequency of prescribed fault conditions. F/A-18 flight control computers processed these inputs and control actuators accordingly. The Integration Test Station provided access to the cross channel data link and MUX data which is supplied to the VMSC. A VAX workstation is used to monitor OFCDS diagnostic output and display the system output. Neural network and fuzzy logic monitors monitored actuator current and position and reported actuator performance which was correlated to maneuver templates.

The system level rule base determined actuator health based on the health status and the conditions under which the fault was reported. Context monitors included hinge moment, excessive maneuver and high temperature conditions. Multiple fault instances of the same type were simulated and a correlation strength was developed. The system module considered this correlation and the various health reports to determine the health status. Additional demonstrations were held using an electrostatic actuator (EHA) on a test bench. Neural networks were used to monitor and report on leakage and pump health status.

Propulsion: The health of nozzle actuators and controls was determined by implementing an engine stall monitor, aerodynamic parameter monitoring, and a neural network model of engine stall margin. The system rule base considered the stall margin and operational conditions in determining nozzle health status. The demonstration also emulated the engine stall's impact on the flight control system. The "soft" failure detected by the flight control health monitors because of a drop in hydraulic pressure due to the engine stall was reported but discarded by the system level module.

Navigation: The health of the gyros and accelerometers were determined. This was a data "pull" in that a "straight and level" flight monitor triggered the health reporting process. This flight monitored looked for periods of straight and level flight over a period of ten to twenty minutes. A regression algorithm was triggered to determine any bias of the accelerometers or gyros from the position and velocity error. The time and quality of "straight and level" determined the confidence level of the health report.

5. ROLE OF KNOWLEDGE DISCOVERY AND DATA MINING TECHNOLOGY

Diagnostics are currently costly to mature, particularly on-line diagnostics and algorithms that support condition based maintenance. As witnessed by the high CND rate, the behavior of a subsystem inflight certainly differs from behavior predicted by design knowledge or on ground testing. A major barrier to on-line diagnostics more sophisticated health management concepts is tuning or maturing the system to provide consistent results. To address this barrier, AODS was designed to bootstrap itself to maturity. Monitors trigger the collection of data based on fault or operational conditions. This data is then used to mature the health models used by the system. Maturation of the models described above require analysis of many real-time data streams with a sample rates of over 100 hz. The approach used here was to use techniques such as neural networks and fuzzy logic which can be developed from examples. This is augmented by knowledge discovery and data mining technologies (Fayyad, 1996) which are used to analyze system behavior, search and identify example learning and test cases for model development. The key is to be able to examine large amounts of data to determine and predict subsystem behavior to a high degree of confidence. Also, the data mining capability can be used to perform analyses to determine the benefit of any prognostics, condition based maintenance and associated maintenance scheduling.

F/A-18 flight and flightline maintenance data was used to demonstrate the use of data mining technologies to identify and characterize subsystem health/faults. A typical flight will contain over 1 megabyte of data. A typical sequence of actions required to prepare the data and perform this analysis is shown in Figure 7.

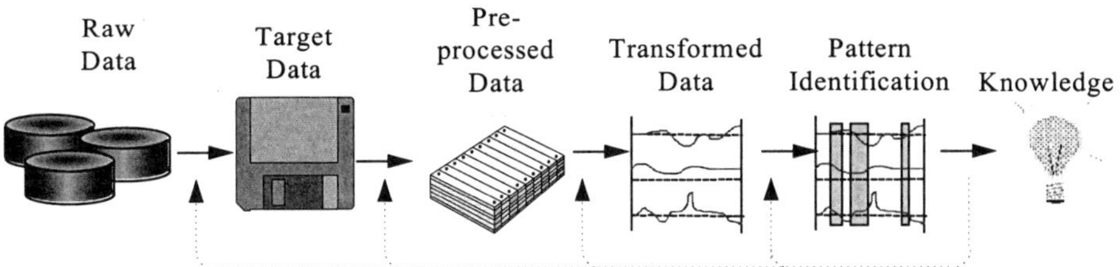

Figure 7 - Knowledge Discovery Process

As an example, several messages in the F/A-18 digital storage unit (DSU) data were identified as the target data. The goal was to characterize the phase of flight from this data and to correlate phase of flight with fault code indications.

Preprocessing: The DSU data was preprocessed to remove all headers, page breaks and footers and then the selected messages which contained aircraft position and fault code indications were extracted. A subset of the data was manually assigned a phase of flight indicator.

Data Transformation: Data transformation consisted of filling in missing records and modifying the time stamp to be consistent across all data sets and rescaling of some of the data entries.

Data Mining: The data mining step consisted of creating a data characterization file or schema in the Silicon Graphics, Inc.—Mineset^{TM} tool suite. Data was viewed and manipulated using such tools as scatter plot, decision tree, option tree, data cloud plot and block diagram (evidence vizualizer). One tool, the column importance algorithm, evaluates all available columns of values and ranks them in order of relevance to a target column (e.g. flight phase indicator).

Interpretation/Evaluation: The column importance tool was used to examine and modify the characterizations of the phases of flight to achieve a desired level a accuracy on the training and test data sets. A rule set was generated to determine the phase of flight based on seven variables. This was used to correlate a type of fault to flight phase. Additional correlations between fault indications and flight conditions were also examined.

Figure 8 shows this process applied to flight data that includes flight control fault indications. Fault occurrences were plotted for the various aircraft tail numbers. Work is progressing on more detailed analysis of fault occurrences and the conditions under which they occur. For example, it is useful to correlate actuator health with the hinge moment limit and analyze bad actor fault sequences and maintenance best practices. This effort will include analysis of subsystem data such as that available on the Cross Channel Data Link (CCDL) between flight control computers. For most platforms this data is output at over 100 hz but this is what is needed to provide in depth insight into subsystem operations.

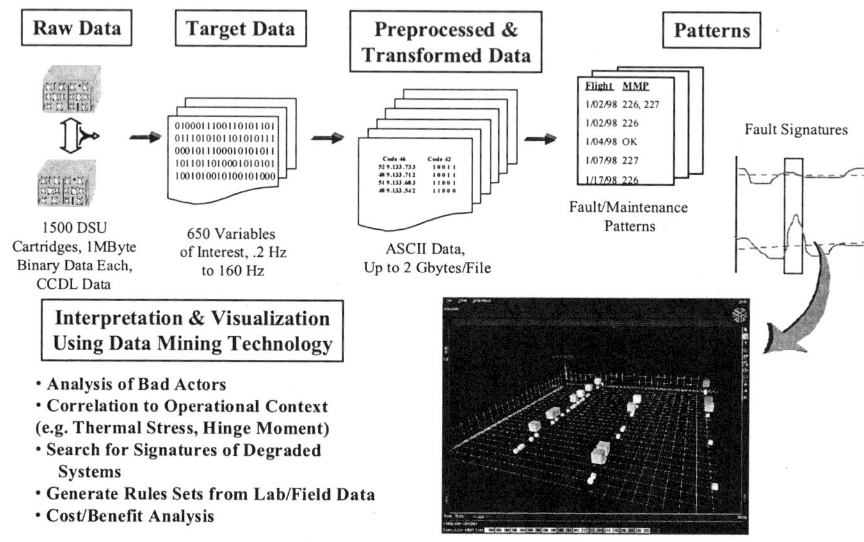

Figure 8 – Analysis of Flight Data and Fault Occurrences

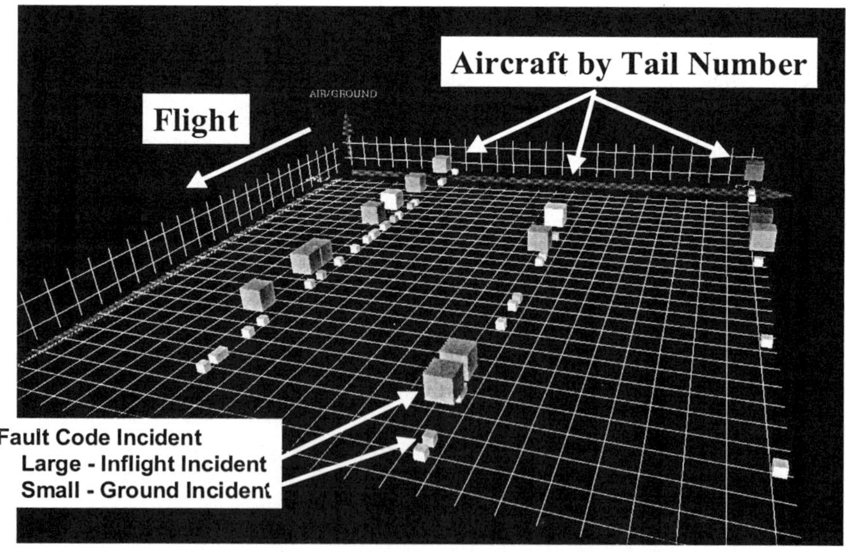

Figure 9 – MineSet ™ Visualization Display of Fault Code Sequences

Figure 9 provides a more detailed view of fault code histories as a function of tail numbers using the Mineset™ visualization tool. Such multiple dimensional color presentations of data is very effective in providing the user with insight into results of the statistical and machine learning analysis tools.

6.   SUMMARY

The AODS research has established a viable architecture and several prototype demonstrations. Recent work has focused on knowledge discovery/data mining to do additional analysis of fault/health of subsystems and of maintenance practices. This work is in turn feeding a cost/benefit analysis to establish the economics of implementing an AODS on various platforms.

7.   REFERENCES

Chiu, Stephen L. Extracting Fuzzy Rules from Data for Function Approximation and Pattern Classification, pp 148-162, *Fuzzy Information Engineering*, eds. D. Dubois, H. Prade, R. R. Yager. John Wiley, 1997.

Fayyad, Usama M., etal, Advances in Knowledge Discovery and Data Mining, The MIT Press, Cambridge, MA, 1996.

Bartz, Darrell O., etal 1997, Knowledge Discovery in Diagnostic Data Bases, Boeing Internal Research and Development Report, December 1997.

Keller, Kirby J. , etal, 1997, McDonnell Douglas Aerospace and Lockheed Martin Controls System, On-line Flight Control Diagnostic System Final Report, Prepared for Naval Air Warfare Center, Warminster, PA; June 1997

Chapter 10

Combining Model-Based and Case-Based Expert Systems

Moshe Ben-Bassat, Israel Beniaminy, David Joseph
IET Intelligent Electronics

Keywords: Model based reasoning, case based reasoning, intelligent diagnostics, service and maintenance support, artificial intelligence.

Abstract: In discussing diagnostic expert systems, the merits of model-based systems vs. case-based systems are often debated. In this chapter, we show that the two are not mutually exclusive. To the contrary, they complement each other. Based on our research with the AITEST diagnostics expert system, we present an approach of combining the two reasoning mechanisms (in addition to other mechanisms) into one integrated system. Depending on the knowledge available, the cases available, and time and cost considerations, the system leaves it up to the user to decide on the relative proportion of case-based to model-based he/she wishes to employ in any given situation.

1.   INTRODUCTION

Diagnostic support software should be evaluated by two critical factor groups (Ben-Bassat *et al.,* 1992): Cost and time to deployment; and accuracy, completeness and efficiency of the diagnostic process. Naturally, there is some tradeoff between the two factors. For quick and less-costly deployment for a non-trivial equipment type, you may have to compromise on the quality of the system's diagnostic support. The question, therefore, is: for a given budget of time and money, which approach produces the best diagnostic performance?

In most practical real life applications, it is more cost-effective to apply case-base reasoning after the system already has some basic initial knowledge of the unit under test (UUT) and its domain. We believe this is true even when available knowledge is relatively limited. We will contrast this to case-base reasoning systems that start from scratch and use only historical cases for making diagnostic decisions for new cases. Such systems will be denoted by O-CBR (O stands for Only). Most commercially available CBR tools are O-CBR tools. Model-Based reasoning systems will be denoted by MBR.

In the following two sections, we present a short introduction to CBR and MBR, taken from a more detailed analysis of different approaches to fault isolation support software (Ben-Bassat *et al.*, 1998).

2.   A BRIEF HISTORICAL PERSPECTIVE

The early days of Artificial Intelligence (AI) research focused on problem solving methods (reasoning processes, inference algorithms) with relatively low emphasis on domain specific knowledge. In the mid 1970s, we began to realize that "if a program is to perform a complex task well, it must know a great deal about the world in which it operates (Lenat and Feigenbaum, 1987)." This is known as the Knowledge Principle.

In the 1980s, research shifted to knowledge-based systems covering knowledge representation and acquisition techniques, and inference algorithms and program architecture that operate effectively on large knowledge bases. Numerous expert system projects started all over the world in just about every field. Many systems were successfully deployed; while many others never matured to a level good enough for practical use.

As practical experience accumulated, it became clear that although the program's performance improved as the knowledge base was enriched, it took a very long time and laborious efforts to construct and perfect the knowledge base before the expert system reached adequate performance. For commercial applications, this brought up the issue of cost-effectiveness. Time to deployment and the cost of development became critical issues.

Case-Based Reasoning (CBR) is not new to the engineering community (Dasarathy, 1990) for a comprehensive review. Its roots are in the classic nearest neighbor classification techniques developed in the 1960s and 1970s for pattern classification, e.g., character recognition. Case-based reasoning was later re-discovered and improved by the Artificial Intelligence community. It is sometimes applied in a less structured way, but the principles remain the same (Carbonell, 1992).

The key principle of the use of CBR in fault diagnosis is that cases that present similar symptoms and findings are likely to result from the same faults. Using this technique, a new case is compared with a set of historical cases of known diagnosis. The "Nearest Neighbors" are identified and the diagnosis of the new case is determined according to the known diagnosis of its nearest neighbors. If there are too many candidate near neighbors, the technique suggests a question or test. The answer to the suggested question or the result of the suggested test is added to the new case and hopefully reduces the number of candidate near neighbors.

CBR is one technique in the rich domain of Machine Learning, which has always been a principal sub-field of AI. As it deals with automatic generation of new knowledge and skills, it could potentially replace part of the manual hand-crafting of knowledge bases, thus improving the cost-effectiveness of building expert-systems.

Up to the mid 1980s, the focus in machine learning research was on the methods. Complex programs were built with fairly rich methods, but with relatively small initial knowledge base. Achievements were not very significant. The programs could not generalize beyond a relatively limited boundary, and were also very "brittle" when reaching this boundary. (That is, they would sometimes come up with a ridiculous conclusion, that would outweigh all other "smart" conclusions, and put at risk the program's credibility for field use).

These attempts in machine learning concluded with fairly uniform consensus in the AI community about the key role of initial domain knowledge for machine learning. In their seminal paper, Lenat and Feigenbaum (1987) state: "Knowledge Facilitates Learning (Catch 22): If you don't know very much to begin with, don't expect to learn a lot quickly"; Or in other words: "The more one knows, the faster one can discover more", or "learning begins to accelerate due to the amount already known". Another somewhat exaggerated way of phrasing this principle is attributed to Marvin Minsky: "You cannot learn anything unless you almost know it".

The implication of the knowledge principle to case-based learning is that it is almost never beneficial to start from scratch. If you load the expert system with some basic initial knowledge, then the amount you can learn from a given set of cases increases markedly.

An O-CBR (Only CBR) approach assumes total ignorance about the unit being tested, so the expert system has to learn everything from scratch. If the knowledge elicitation starts with even a rough model of the unit, the cases will become more meaningful, and the computer diagnostic recommendations will be clearer. (If you have an inference algorithm to operate on the UUT model).

The underlying assumptions behind these claims are that loading some useful initial knowledge base takes shorter time than accumulating the cases, and that the knowledge-based diagnostic algorithm can indeed learn more than we can learn by simple matching to historical cases. Before demonstrating these claims, let us clarify them further by using medicine and human learning as an analogy.

3. ANALOGY TO MEDICINE

For many years, we were ignorant of the anatomy/physiology of the heart. Physicians could only use reference to historical cases as a basis for diagnosis. Today, we know fairly well the structure and functions of the heart, and thus today's physicians, familiar with the anatomy and physiology of the heart, accomplish much more than past physicians, who relied only on historical cases.

In medicine, the "engineer" who designed the "unit" under test (the human body) did not leave any engineering drawings behind. Thousands of years of medical research are aimed at reverse engineering in an attempt to discover the "unit" model. This research is oriented towards eliciting models out of case-based information. Once a model is discovered, complete or partial, it essentially encapsulates in compact structures the knowledge of the many cases that led to it. The cases continue to serve as a reference, but the whole model is worth much more than its individual parts. Most importantly, teaching the model takes much shorter time than collecting the individual cases that led to it.

In diagnosing equipment units, we rarely find ourselves in a situation of full ignorance. In most real life applications, there exists a certain basic amount of knowledge. For instance, a good draft of a parts list of the smallest repairable / replaceable units (SRRUs) and a draft list of symptoms and tests that may be observed or measured are always available. We may not have full comprehensive lists when we start, but, on the other hand, we rarely have zero. This places a disadvantage on any tool that builds this list from zero, on the fly, as you collect cases in the field.

O-CBR takes the black box approach assuming full ignorance about the UUT (Unit Under Test) and about the UUT-domain, e.g., electronics, hydraulic, etc. This ignorance limits the applicable reasoning methods. On the other hand, model-based learning can integrate the impact of a single case or a group of cases into the unit model, thus achieving a much stronger diagnostic vehicle.

The analogy to medicine also teaches another important lesson: If we ignore available knowledge, that knowledge will gradually disappear. Once knowledge about the UUT is gone, a disproportionally large effort is required to regain the knowledge through reverse engineering. Sadly, such wasted effort is far too common in many situations where service engineers have no access to design knowledge. These engineers are forced to learn how to service the UUT through case-by-case experience, slowly and painfully. Even then, their knowledge is experience-based and very difficult to transmit back to the designers in order to improve the product quality, since the knowledge of the design engineers is based on their understanding of how the UUT works - in other words, the design knowledge is model-based.

4.   CASE-BASED REASONING

In CBR (Dasarathy, 1990; Carbonell, 1992), many cases are stored, each of which is composed of a description of initial findings, of questions asked (or actions recommended) by the expert, the answers to these questions (or the results of these actions), and the final diagnosis. Figure 1 shows two such cases for a hypothetical photocopier example.

CASE 1: *Initial finding*: Photocopy shading quality low.
Question: is glass cover clean?
Answer: Yes.
Question: Is quality of test pattern OK?
Answer: No, dark areas quality low.
Diagnosis: Electronics board failure

CASE 2: *Initial finding*: Photocopy shading quality low.
Question: Is glass cover clean?
Answer: Yes.
Question: Is quality of test pattern OK?
Answer: No, both light and dark areas quality low.
Diagnosis: Camera failure.

Figure 1 - Two CBR cases for a hypothetical photocopier

Once an initial finding is reported, the CBR software scans for matching cases and selects a question to ask in order to differentiate between the currently matching cases. The answer to that question reduces the number of matching cases selected in the next step of the fault isolation process, until the fault is found or no more cases match the findings. For instance, when the photocopy shading quality is low and the glass cover is clean, both cases in Figure 1 match the findings. The software can then select the question "Is quality of test pattern OK?" The answer should leave one possible case, so we can report the diagnosis associated with it.

CBR knowledge is usually gathered by monitoring the expert as he/she isolates faults. This can be an advantage when the expert cannot devote enough time to building the knowledge. Additionally, CBR does not require the expert's knowledge to be expressed in a formulation that a computer understands, and therefore does not requires much training. Instead, it works by gathering sequences of questions and answers. However, when going beyond small "toy problems," a large number of cases become necessary, and CBR practitioners have to carefully plan and maintain the structure of the case base.

5.   MODEL-BASED REASONING

Unlike Case-Based Reasoning, which represents its shallow knowledge as a collection of cases, Model-Based Reasoning represents its deeper knowledge as a model of the cause-effect relationships governing the equipment and the interaction of its possible functions malfunctions with its observable behavior.

There are many possibilities to define such models. A good introduction is in Simpson and Sheppard (1994). In Ben-Bassat *et al.* (1998), we made a broad distinction between physical modeling and fault modeling. Physical modeling (Bobrow, 1985), explicitly models the actions of each component and each of the interactions between components. Such a model is powerful enough to simulate (at least qualitatively) the behavior of each component. Some tools model only the behavior of non-faulty components. Other tools extend the simulation to also model the behavior of faulty components. On the other hand, fault modeling is not concerned with how and why the equipment works, and focuses on the possible ways that the equipment can fail.

The main idea in fault modeling is that we do not need to know exactly how everything works in order to understand how to isolate a problem. Instead, fault modeling specifies which functionality of the equipment (as observed by symptoms and measured by tests) would be affected by which of the possible equipment malfunctions. Therefore, such methods are sometimes referred to as "Dependency Modeling". In AITEST and TechMate, the list of elements whose malfunctions that may cause symptoms or test failures is called "test paths" to emphasize its close relationship to paths of the signals flowing through the UUT during this test. Figure 2 shows the test paths for our photocopier example. The fault isolation proceeds by listing the suspected faulty components, then trying to raise or lower their "suspicion level" by choosing tests that separates them into hierarchical categories. For example, after symptom S1 is reported, all the components on its test path are suspected. The software now looks for a test whose path contains only a few of the suspected objects, say test T3. Suppose that the service engineer reports that test T3 yields good reproduction in at least one exposure setting. The modules on the path of T3 are then assigned a lower suspicion level, but not totally eliminated–the bulb might still be faulty in not emitting enough light, for example. This process continues until only one suspect is left, or the list of suspects cannot be narrowed down. In that case, the software will look up which repair action is associated with the suspects, and suggest the quickest/cheapest one.

Path for symptom S1 (bad reproduction quality): electronics board (exposure control); bulb; glass; toner supply; camera
Path for test T1 (power-on self test): electronics board, toner supply (toner level), motor (response to commands)
Path for test T2 (Copying test-pattern original, automatic exposure) Depends on specific failure mode:
Dark areas quality low, light areas quality OK: toner supply; electronics board (exposure control); camera
Dark areas quality OK, light areas quality low: glass; bulb; electronics board (exposure control); camera
Path for test T3 (Copying test-pattern original, manual exposure): glass; bulb; camera

Figure 2 - Fault model for a hypothetical photocopier

Modeling similar to the method described here is one of the foundations of the IEEE 1232 AI-ESTATE (Artificial Intelligence Exchange and Service Tie to All Test Environments) international standards (IEEE, 1995; IEEE, 1997; IEEE, 1998; Maguire and Sheppard, 1996; Orlridge, 1996).

In our research with the AITEST expert system, we concentrated on fault modeling, because it can achieve high accuracy even with a limited amount of knowledge (Ben-Bassat *et al.*, 1998), and also lends itself better to the integration with CBR explored in this chapter.

A good fault model may be used to solve any problem, whether the human expert considered it explicitly or not, as long as the faulty component and the tests and actions required are modeled; and each part of the model can be used for solving many different symptoms, making the required knowledge more compact and easier to manage.

Note that model development requires a high level of understanding of the equipment, since it needs to take a global view. In some applications, it might not be easy to dedicate the time of an expert with such high-level understanding to this task. Such understanding is not always necessary with CBR methods. Compared with physical modeling, knowledge accessibility is usually much less of a problem for fault modeling, since the amount and complexity of required knowledge is usually much lower. In the photocopier example, a fault model just needs to know that low toner levels reduce the quality of dark areas, without modeling how toner particles are charged proportionally to light level, transferred to the drum and heated to bond to the paper. On the other hand, the physical model is deeper, and therefore its treatment of especially obscure situations could be better than that of fault modeling.

6.   RELATED WORK ON CBR AND MBR FOR SYSTEM TEST AND DIAGNOSIS

In this chapter, we refer to the "pure" form of CBR, as described above. While there have been many advances over this pure form, it can be seen that these advances are close in nature to the method presented here: combining the inductive learning with analytical reasoning. In these methods, usually falling under the ILP (Inductive Logic Programming), the analytical reasoning essentially extracts some sort of model out of the stored cases, as in the FOIL expert system (Quinlan, 1990) and its later extensions, such as the FOCL expert system (Pazzani and Brunk, 1992); and in the GOLEM ILP system (Muggleton and Feng, 1990). More recent studies along these lines include Bearse (1998) and Sheppard (1998). However, these methods, unlike the method presented here, do not start with knowledge about the problem domain (in this case, isolating faults in equipment), and, therefore, require a large number of training cases in order to reverse-engineer this knowledge. The method presented here builds upon problem-domain knowledge, as well as some amount of initially entered knowledge about the specific equipment being repaired. Then, it uses the initial knowledge to efficiently structure the model being extracted from accumulated evidence. An important result of this approach is the transparency of the generated model–it uses standard engineering terms, relationships and diagrams instead of less approachable formalisms such as Horn clauses. This transparency improves communication and collaboration between the domain expert (usually not trained in formal logic methods) and the expert system.

7.   THE POWER OF MODEL-BASED REASONING

Using Model-Based Reasoning (MBR), the diagnosis proceeds at a deeper level of understanding. This means that a small, coherent knowledge base can provide a high level of efficiency and accuracy in the diagnostics process.

An engineer who understands the UUT can usually infer quite quickly which module is the most likely cause for a certain set of symptoms and test results. To make such an inference, the engineer usually needs to consider just a small, localized part of his/her knowledge of the UUT. On the other hand, an O-CBR system has to compare the given set of symptoms and test results to all recorded cases. There are bound to be a few recorded cases which have a statistically high level of similarity to the current case, but a knowledgeable engineer will easily discard most of these cases since they are not at all similar when you consider how the UUT actually works.

The main argument against MBR is that in some situations, there is not enough available knowledge or not enough time to construct a model that is rich enough to support efficient and accurate diagnosis. This provides us with the incentive to consolidate MBR with CBR in order to gain the best of both worlds. Our research along these lines resulted in the AITEST-CBR equipment diagnosis expert system, continuing our earlier research on the AITEST expert system (Ben-Bassat *et al.*, 1988). The results of this research, which we present in the following sections, are very encouraging, and have lately been incorporated into the new generation of AITEST, renamed TechMate.

8.　MODEL-BASED REASONING IN AITEST AND TECHMATE

In AITEST, and in its newer generation expert system TechMate, the fault model knowledge is basically organized in several parts:

1. A hierarchical list of subsystems, assemblies, boards, cables etc. down to the SRRU (Smallest Repair/Replaceable Unit) level appropriate for the specific UUT and service/maintenance environment.
2. (Optional) The different failures that can occur in the equipment: a list of malfunctions that may happen in each SRRU.
3. The different symptoms that may be observed, and the different acceptance tests and diagnostic tests that may be performed to narrow down the diagnosis. Optionally, each symptom and test may be subdivided into parameters (e.g. for a wave shape test, there may be separate parameters for pulse width, rise time and fall time) and failure modes: each test or parameter might fail in ways that differ in their diagnostics implications (e.g. a voltage test that fails in the "too high" mode will often assign lower blame to the power supply compared to a "zero" failure mode, if the power supply is more likely to burn out than to be out of calibration).
4. The relationship of each of these failures to the symptoms that can be observed and the tests that can be performed:
   - "Test path": The list of SRRUs that are related to the symptom or test, i.e. should be blamed to some degree of the test's result is Pass, or exonerated to some degree if the result is Fail (Figure 2). The top three arrows in Figure 3 mean that if Symptom 1 is observed, we should suspect A2, B1 and B2.
   - (optional) "Blame distribution": The elements in each test path may each be assigned a different blame, which may be differ for the various test parameters and failure modes (see the bottom three arrows in Figure 3, which mean that if test Acc2 fails, we should assign 50% of the blame to board A1, 30% to board B1 and 20% to harness B2).

- (optional) "Test Relationships": For each type of SRRU malfunction and each test (or other members of the test hierarchy—symptom, parameter, failure mode) the knowledge may specify the probability that the occurrence of that malfunction will cause that test to fail (see Figure 4, which specifies that if malfunction MF3 occurs in Board A2, then there is a Very High (VH) probability that Test Acc1 will FAIL and a High probability that Test Diag2 will FAIL).

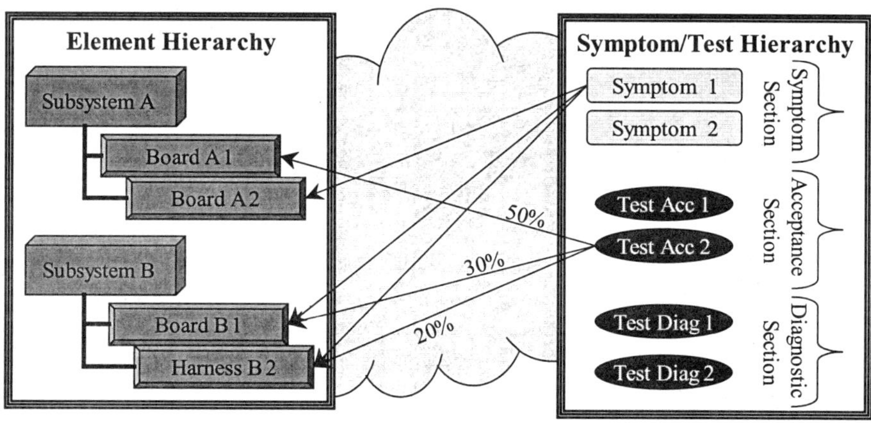

Figure 3 - TechMate diagnostic model: Test paths with blame distribution(bottom three arrows) and without blame distribution (top three arrows)

9. WHERE CAN MODEL-BASED KNOWLEDGE BE OBTAINED?

At first sight, one would suppose that building a model-base requires deep, rare understanding. This may be true if the model-base will be the only source of knowledge. However, if we plan on filling-in missing knowledge using CBR, then the problem becomes much simpler: How do we obtain a limited starting point around which the expert system will automatically classify and structure case data?

In fact, such knowledge is usually easy to find. Surprisingly, it is often available for automatic import, if you know where to look: Element hierarchies may be obtained from CAD (Computer-Aided Design) data and ERP (Enterprise Resource Planning) product trees. Test hierarchies may be available in maintenance procedures. Additionally, even for service and maintenance, tests are often possible to adapt from TPS (Test Program Sets) used in final integration and testing in the equipment's production line. If reliability analysis has been performed (e.g. using FMECA or FMEA methods), a fault modeling system can import it to obtain large parts of the needed knowledge about elements, tests and element-test interactions (see Figure 5). A similar import can be performed using simulation data (Fenton and Webber, 1993).

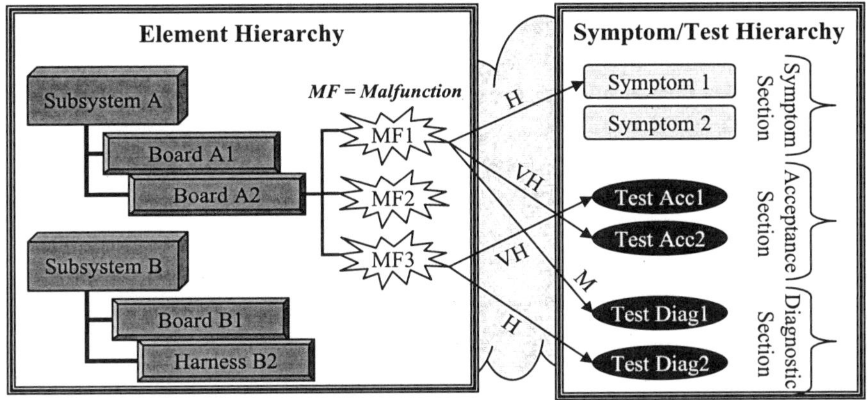

Figure 4 - TechMate diagnostic model: Test relationships
(VH - Very High; H- High; M- Medium)

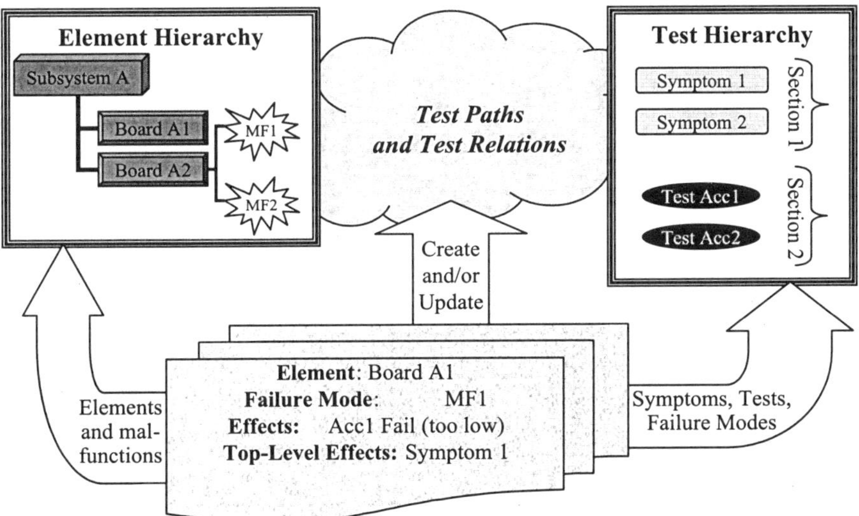

(Sources: FMECA / FMEA Analysis; Fault Simulations)

Figure 5 - Importing fault models from reliability analysis and fault simulation

10. INTEGRATING ANALYTIC REASONING (MODEL-BASED) WITH ANALOGICAL REASONING (CASE-BASED)

Three main types (layers) of knowledge may be identified for equipment diagnosis. AITEST-CBR utilizes all three of them in a integrated fashion (Figure 6):

- Universal "school/textbook" knowledge on diagnostic considerations and processes, e.g., how to use topological (structural) knowledge.
- Specific knowledge about the UUT in terms of structure, function, and behavior under the tests defined for it, e.g., relations between symptoms and faults.
- Historical cases representing experience with the UUT.

AITEST contains a built-in universal knowledge base about equipment diagnosis, and templates to enter UUT-specific knowledge. The latter include the UUT list of modules. The list of symptoms and tests, and the "blame distribution" of failing tests (symptoms) over their related modules (what we call "test relations matrix") and includes options to enter top level (or detailed) block diagrams and the path of each test through the modules and links in the block diagram. Equipped with the unit topology and the test paths, the system derives automatically the relative diagnostic weights of a given test or a set of tests with respect to all the relevant modules. This automatic derivation drastically reduces time to deployment.

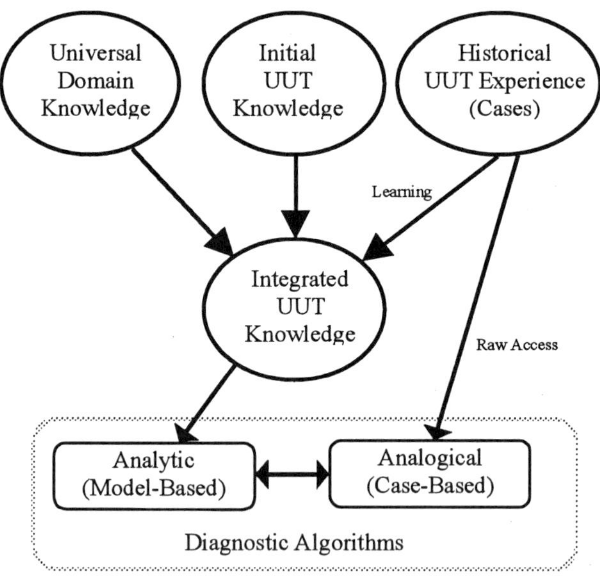

Figure 6 - AITEST-CBR Integrates Universal, UUT-Specific and Historical Knowledge

The user interface of the system's knowledge acquisition tools communicates in engineering terms only and there is no need for a knowledge engineer mediator. The translation into internal "AI data structures" is done automatically. Internally, the knowledge base is primarily represented by structures conceptually similar to probabilistic inference networks, and are complemented by frames and rules. The weights in the test relations matrix may be viewed as conditional probabilities.

The algorithm for diagnostic assessment of a given set of test results is implemented as an evidence propagation process along the network representing the relationships between the tests and the unit components, sub-assemblies, links, buses and harnesses. This process is further controlled by the knowledge derived from the frames and rules that cover more refined specific situations. The test sequencing algorithm is goal-oriented. For a given goal, it sequences the tests by balancing their information content (measured by entropy-type functions) with their execution "cost" and time. Presented with a new case, the systems diagnostic algorithm operates on the integrated universal and UUT-specific knowledge base for assessing the fault likelihood of the various modules for any given set of test results. (As opposed to an O-CBR system which is quite limited when confronted with a case that it has never seen before).

AITEST-CBR also contains a data base where you can store pre-prepared historical cases and cases accumulated during your work sessions. This case-base is approached at two levels:

10.1 Learning: Integrating the Cases Into the Diagnostic Model

AITEST-CBR integrates the knowledge included in the cases into its current version of the UUT-specific knowledge-base. This is done in several ways, including updating the failure rate of the unit modules, adjusting the weights that relate test results to modules, or "discovering" that a module has several fault modes, each of which corresponds to a different function of the module. In the latter case, the system will automatically add a new column to the matrix and generate the weights for this column. Depending on the size of your case-base, and your assessment of the validity of the current model-base, the system lets you choose between accelerated learning where the impact of the cases on the knowledge base will be more significant, and moderate learning where the existing knowledge base is adjusted more conservatively.

The result of the integration of cases into the diagnostic model is a better, more accurate UUT model, reflecting newly deduced knowledge about the UUT. This new knowledge is expressed in model-based terms, e.g., "Analysis of cases gathered over the last month indicates that the power supply can fail in such a way that most of the UUT functions correctly, except that output 3 has high noise levels". Such knowledge is used to fix subsequent failures, and is also in the best form for presenting feed-back to the design engineers, unlike the raw case knowledge.

Note that this automated learning can be supervised by a human engineer. This can help prevent irrelevant or out-of-date information from affecting the diagnostics knowledge. Moreover, the human engineer can instruct the automated learning to "forget" cases that are no longer relevant.

10.2   "Conventional CBR" - Direct Retrieval of Similar Cases

For any given case, AITEST accesses the case-base at the raw level, and displays the similar cases previously seen - or simulated - together with their final actual diagnoses—just like all other CBR tools. This, of course, can be done at any intermediate stage of the diagnostic process with any number of symptoms and test results. In this respect, AITEST-CBR does whatever O-CBR does, and much more.

The use of direct raw level case similarities is particularly useful for difficult situations where clear cut diagnoses cannot be reached. In these situations, looking at previous cases may contribute the last piece of evidence required to make the final decision. In addition, sometimes, the technician may think differently than the expert system ("gut feeling"). Looking at similar historical cases may be helpful in resolving differences in opinion between the expert system and its user.

AITEST-CBR makes use of the historical case-base to deal with exceptions, an issue of major importance in diagnostic expert systems. For any given new case, it scans the case-base for similar cases and compares its "analytic" diagnosis with the diagnoses of similar cases. In case of differences it will call the user's attention to the deviations. If the case-base includes historical cases that represent the exception rather than the rule, then this mechanism will strongly assist in dealing with exceptions.

The following examples illustrate a variety of situations where an O-CBR tool requires a large number of cases to learn a fact that can be easily given to AITEST, and also some situations where most O-CBR tools do not have the expressive power to receive and use important knowledge for efficient test sequencing.

Example 1–A small amount of initial knowledge replaces many examples. Consider the ignition/injection system of a car that includes the Electronic Control Unit (ECU), a set of sensors; e.g., air pressure sensor, and the functional modules: main panel, distributor, plugs, etc. There is a group of tests T_1, T_2, ... T_8 designed to verify the functionality of the various individual sensors. All of these tests also involve the ECU. If one or two of these tests FAIL and the rest PASS, it makes more sense to place the blame on the relevant sensor(s). However, if all of them FAIL, it is very likely that the ECU is faulty, because the only other alternative is that all of the sensors fail at once and this is not too plausible.

A general rule leading to this conclusion is included in AITEST Universal Knowledge Base, and it may be further tuned to capture more refined knowledge using the system's tools for UUT-specific knowledge. On the other hand, many cases are required for an O-CBR tool to arrive at such diagnostic conclusions. In this case, the fact that is evident to the domain expert, and easily elicited using simple diagrams, would require many examples for any learning expert system that starts with no knowledge.

Example 2–Modeling cost factors and safety rules. Many real-life considerations are difficult to state or learn without the explicit statement of some initial knowledge. For example, even if examples include the time required for each test, and even if the learning tool is programmed to learn these times, it would take many examples to learn the following fact. "The time to perform Test3 depends on whether the box is already open, as a result of earlier test actions Test5 or Test11." Other important information easy to elicit into a model, but hard to learn–and in this case, possibly too dangerous to leave for learning–is the rule "If Test11 failed, do not run any other test before verifying correct electrical grounding at point TP6 (Test59)." Most or all O-CBR tools do not allow a direct expression of such rules. Furthermore, some O-CBR tools will skip a test such as the Test59 if they determine that in all the cases under consideration, Test59 had the same result–correct grounding. Generally, this is a good way of eliminating redundant tests and getting faster diagnosis. However, in this specific example, skipping Test59 when Test11 failed may be dangerous.

Example 3–Ignoring redundant tests. Consider a car engine with following historical cases:

<table>
<tr><td>Historical Case 1:</td></tr>
<tr><td>"Mass Air Flow" Fail</td></tr>
<tr><td>"Volt Pin 7" Fail</td></tr>
<tr><td>Problem: Module "Control" is faulty</td></tr>
</table>

<table>
<tr><td>Historical Case 2:</td></tr>
<tr><td>"Volt Pin 7" Fail</td></tr>
<tr><td>"Volt Pin 2" Pass</td></tr>
<tr><td>Problem: Cable (Line) 26 is faulty</td></tr>
</table>

Now, let us consider the following case:

New Case 3
"Volt Pin 7" Fail
Problem: ?

An O-CBR system will find a match with both cases: 1 and 2, therefore it might offer both tests "Mass Air Flow" or "VoltPin 2" as if they were of the same importance. However, hardly no technician would perform "Mass Air Flow" after "Volt Pin 7" is already known to fail. Referring to wiring diagrams, we can see that "Mass Air Flow" is a more general test, that tests the Control unit, and many other links that go into the Mass Air Flow module. What would the result of "Mass Air Flow" test add to refine the diagnosis? AITEST will not offer "Mass Air Flow" in such a case.

Note that for the reasons mentioned in Example 2, many O-CBR tools do not take advantage of redundancy even after enough cases have been gathered to indicate that such redundancy exists.

11. COMPARING O-CBR TO COMBINED MBR AND CBR IN DEVELOPING KNOWLEDGE

If an O-CBR system is to be based on real life authentic field cases, the only situation when sufficient knowledge can be developed quickly is in UUTs with very high frequency of failures, and in which the faults are well distributed over all of the unit parts. (In this situation, there may be worse problems than the need for a diagnostic support tool.). In all other cases, we will have to wait a long time for a sufficiently large and representative set of cases, before the expert system learns enough to yield adequate performance.

An authentic case obtained from a good field engineer is likely to contain just about the minimal set of tests sufficient to isolate the existing fault. Many of the test results may not be available because they were not needed to diagnose the case. The field engineer may also have noticed several things which were crucial in the decision making process, but the stored case may not include these observations. It's long been recognized by the Artificial Intelligence community that unstructured knowledge acquisition can lead to incomplete and inconsistent knowledge inside the expert system.

The alternative to real life cases are simulated/theoretical cases. Two options are then available: Build a software simulation model of the UUT and use it to generate a large set of simulated cases (Fenton and Webber, 1993) or to generate simulated cases manually.

For complex large units, building a simulation model is either impossible or cost-prohibitive. In a quantitative model, available from the authors, we show that the effort in manually creating the simulated cases required for adequate training decreases sharply with a comparably smaller investment of effort in eliciting the basic, accessible knowledge from the domain expert.

CBR works best only if the structure of the generated case base is carefully designed, and this design, together with a good understanding of the UUT, must be followed strictly while defining the training cases. If we have decided to hand-craft the case base, we may as well spend some time on teaching the expert system more general principles.

In many O-CBR tools, cases need to be of similar size and structure. To see why, consider two cases: One with just two questions, and one with ten questions which include the same two questions and answers as the smaller case. Most O-CBR tools will hit the smaller case first, and will either avoid suggesting the second case, or suggest it with a low probability; the smaller case is a much better match. In many situations, this behavior does not fit the users' expectations. To solve this, many O-CBR users have established rough guidelines for contents and size of each entered case.

What do these guidelines mean? Actually, they convey the engineer's understanding that the two questions were not really enough to eliminate all but one of the possible suspects. In other words, the engineer entering the case is using her understanding of the equipment to add more questions and thus complete the case. This is Model-Based Reasoning, applied without the benefit of software support, to create material for Case-Based Reasoning.

This is the central claim of this chapter: A partial UUT model is always available. Depriving the expert system of access to that model can only result in inefficiencies. If the model is too partial to allow efficient diagnosis, it can be supplemented by additional cases. With time, these cases will improve the quality of the UUT model.

It is possible that due to lack of time or knowledge, the initial model will include only lists of modules and tests, without any indication of the effects of different module malfunctions on the test results. These lists do not contain enough knowledge to make any diagnostic assessments. The result is that, initially, the combined MBR/CBR expert system will be driven by raw access to the case-base. However, the MBR learning algorithms are able to generalize the gathered case histories and automatically build a model-based representation of the UUT. In turn, the MBR diagnostics algorithms access this generalized model to generate diagnostic assessments even for cases that are very different from any cases seen before. This process can quickly reach a stage where most of the diagnosis is generated by AITEST's inference algorithms, and only exceptional cases require raw access to the case-base.

Additional benefits of this MBR/CBR combination include:

- The model base can automatically structure the cases into a consistent form, ensuring compliance with safety and logistics rules, and assuring consistent treatment of large and small cases alike.

- Comparing the model to the new cases can highlight areas where seemingly inconsistent information has been given, and highlight areas where more cases are required to complete the model.

- The model can be directly used to generate feedback on the design's design testability, maintainability and reliability. Since all of these disciplines are model-based, using CBR for diagnostics makes feedback quite difficult.

- With O-CBR, it's difficult to react quickly even to small design modifications: In the absence of knowledge linking the design modification to its effects on faults and effects, updating requires a long, manual, error-prone update process. When the knowledge is centered on a model, the new changes can be defined in one place (e.g., modified signal routing), and all the affected cases will be identified and modified automatically. In many situations, even with no explicit manual model modification, the model will detect the changes after several relevant cases appear, and update all other relevant cases in its case base.

12.   CBR CASE STUDIES SHOWING THE NEED FOR PRUNING AND STRUCTURING THE CASE BASE

To teach an O-CBR expert system, the bare minimum required is one case for each possible malfunction of each replaceable element (and many more when we consider that the same malfunction might cause several different symptoms and failure patterns). Additionally to the large number of required cases, real-life considerations require imposing structure on these cases in order to overcome the weaknesses caused by lack of deeper modeling tools.

PC AI (1997) reports a CBR study for diagnosing a jet engine. The team gathered 23,000 troubleshooting reports from a period of 10 years. In eight months, a first version was developed using this data, but the diagnostics it yielded were too fuzzy, due to incomplete and sometimes erroneous raw data. To improve the diagnostics performance, the case data had to be cleaned and pruned, involving the review of all cases at a rate of 15 cases per hour—fast indeed, but still translating into many work hours. The cleaning process resulted in 1500 "clean" cases, and the team then added cases generated from the manufacturer's manual to ensure no "holes" were left in the case database.

In another study (Thomas *et al.*, 1997), a team developing a CBR-based tool for a large consumer electronics support center found that it had to define a structured procedure to acquire and structure case data to achieve a common structure, and the implementation stalled due to difficulties in getting case material from the experienced call center engineers, who initially felt that "AI insulted their intelligence". It is interesting to note how the team handled this obstacle: They asked the experienced engineers to provide opinions on contents and structure. This amounts to asking the experts to shift to a deeper knowledge level, and then using the deeper knowledge to enhance the knowledge maintained in the shallower CBR form. While this works, it begs the question of whether the whole knowledge elicitation, management and deployment process could benefit from a more rigorous shift into deeper representations. It seems that the experts find it easy, but trivial and boring, to recite individual cases, but are far more interested in the principles that generate and classify the cases. These principles are directly related to the equipment's design, technology, structure and principles of operation. It is our claim that therefore the principles are best expressed in the form of a model.

While the study reports how these problems were resolved, it clearly indicates that CBR methods do not yield instant results, and require a strong commitment as well as resource allocation. This is not shameful, but it does contrast with the common perception that "all you need to do is collect the cases while working, and let the software deal with the rest".

13. HANDLING LESS COMMON FAULTS: THE FALLACY OF THE 80-20 RULE

An O-CBR approach is sometimes justified by the following argument: "80% of the cases result from just 20% of the potential faults. Therefore, it is enough to have a support tool that handles the 20% most common faults." However, a contradictory claim based on the same 80-20 rule might be more correct: "Most service engineers know the 20% most common faults, and repair such faults very quickly. The less-common faults take a long time to repair. Therefore, 80% of the service time and costs are due to the less-common faults."

Deploying diagnostics software that only handles the most common faults has two implications: First, the performance improvement achieved by supplying service engineers with such software would be low, since even novice service engineers already know the common faults and deal with them effectively. Second, the time required to diagnose a failure would have a high variation: the common cases would be diagnosed quickly, but there are 20% of the cases which the diagnostics software cannot support. The costs in time, repeat visits, spare parts usage, and customer dissatisfaction would be very high—far more than you would expect from the lower frequency of such cases. A quantitative model available from the authors explores this more fully. A very conservative application of this model shows that an expert system that "understands" 80% of all expected cases will raise service engineer productivity by only 14% compared to working with no expert system at all, while an expert system that handles 95% of all expected cases will yield an improvement of 64%. With O-CBR, the investment required to go from 80% to 95% is far too high, since it requires explicit representation of a large number of rare cases inside the case base, but adding each rare case alone does not translate into any measurable improvement.

While reviewing this chapter, the editors have commented that even the assumption made above—that the 80-20 rule is indeed correct and that therefore handling 20% of all possible faults is equivalent to handling 80% of all expected cases—is being invalidated by recent trends in reliability engineering. Reliability engineering uses statistical methods to analyze the most common faults and find ways to prevent them. The end result that while the total number of faults per time period is reduced, their spread over the possible causes becomes more even: In a sense, there are fewer "common faults" and more "rare faults". If the 80-20 rule is thus converted into a "60-40 rule", focusing on common faults becomes an even less attractive option.

14.   ANALYZING THE TIME REQUIRED TO BUILD A CASE BASE

Whatever the number of cases, let us analyze the process of building one typical simulated case pattern and estimate the time required:

Step 1	Select a module MX for which a case is to be built.
Step 2	Identify the possible fault modes of module MX: MX.1, MX.2, MX.3, ...
Step 3	For a given fault mode MX.i, identify a set S of BIT/symptoms/tests that are relevant for MX.i.
Step 4	For the relevant BIT/symptoms/tests, decide on the expected typical result if fault MX.i is present. (There are also going to be non-typical results which are still possible, and this brings us back to the issue of probabilistic patterns).
Step 5	Add to the set S symptoms and tests that are required because of the imposed structure (see the next section), safety and logistic rules.

At the end of Step 5, we have a draft for a simulated case for MX.1. For quality assurance purposes, it is highly desirable that this case pattern be submitted for a review by an additional expert and/or be validated by actually experimenting with the UUT (if this is not too costly). Experience so far with building simulated cases reveals that, quite frequently, the first expert forgets important BIT, symptoms or tests, and several iterations are required before converging to the final version.

Moreover, once we establish a case base for a group of fault modes, it frequently happens that for different fault modes, e.g. MX.6, MY.2, we get the same case patterns. Then, the question arises: do we have a fault coverage problem; i.e. there are no tests that can distinguish between MX.6 and MY.2, or is it an oversight of the experts and further refinement is required? This also takes some time.

We assess that by the time a UUT of a reasonable size and complexity is ready to be deployed, the total time invested will lead to an average of 1 to 2 hours for each fault mode. This includes the first draft, a review by the second expert, group discussions, and quality assurance of the case patterns. Taking 1 hour as average and 1200 fault modes (assuming a moderate-complexity UUT with several hundred replaceable parts, and a few possible fault modes per each part), we get 1,200 hours for the above example before we are ready to deploy the O-CBR tool.

15.   USING A MODEL-BASED APPROACH WITH THE SAME BUDGET

Some people would argue that they are willing to compromise performance for a certain portion of faults in order to shorten deployment time and cut cost and therefore they do not really need 1,200 hours. Whatever budget you allocate, AITEST-CBR is a better alternative to O-CBR. Assume that you allocate a given budget; say 600 hours for deploying an expert system. Using AITEST, you may use 50 hours to build a rough model of the unit and its test, and then use 550 hours to collect 550 cases; real or simulated. If you now apply AITEST-CBR to these 550 cases, the overall benefit is likely to be much higher than using 600 cases over zero UUT knowledge. Moreover, if UUT knowledge is readily available, spending 200 hours to build a more refined model of the UUT and its tests, followed by 400 hours to collect 400 cases would lead to even better results.

In fact, once you perform Step 1 to Step 5 above on your way to manually generate simulated cases, you are already not too far from having the UUT-model required by AITEST, and, if so, you may as well use AITEST's diagnostic engine. In other words, if you do not have a large authentic case-base, then an O-CBR tool with manually generated cases is the least cost-effective alternative.

16.   SHORTENING TIME TO DEPLOYMENT

To summarize the discussion of trade-off between MBR and CBR, see the following figures. Figure 7 shows the performance of a typical Only-

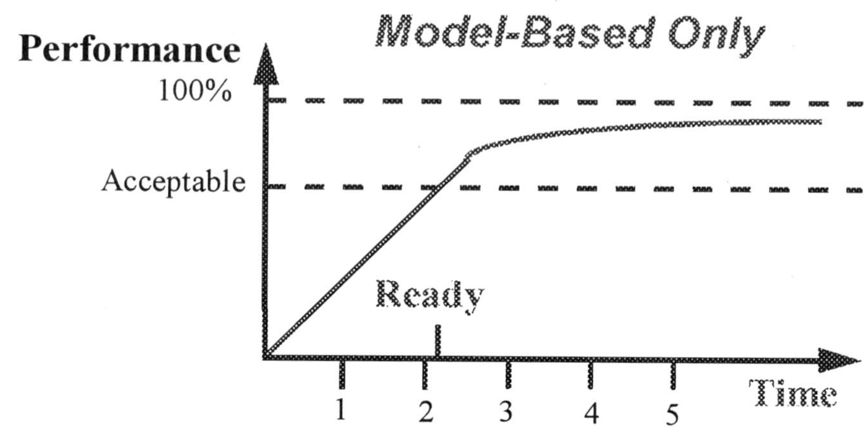

Figure 7: Performance growth for an Only-MBR Diagnostic Expert System

MBR expert system as a function of time, beginning at the start of deployment. The expert system will be considered ready for deployment when the performance passes the "acceptable" performance threshold. The time units are arbitrary and only intended for comparison.

For our purposes here, we may define "performance" as any combination of factors such as accuracy, efficiency and compliance. *Accuracy* is the ability to recommend the correct repair action. *Efficiency* is the ability to reach the conclusion in minimal time, where the time is a function of the number of suggested questions and tests, and the time it takes to answer the questions or perform the tests. *Compliance* is the ability to suggest test and repair actions in a way that complies with all safety and logistics rules.

Performance for only-MBR may only rise via explicit additions to the model knowledge. Figure 7 shows that performance for an only-MBR expert system rises slowly as knowledge is manually entered into the system, leading to some delay until the expert system is ready for deployment. However, the performance level can eventually rise very high, since adding one fact can usually improve the handling of many different cases.

Figure 8 shows the typical performance over time of AITEST-CBR and Only-CBR expert system. Initially, performance rises very quickly. However, as many real-life situations have shown (see above), the performance rise starts leveling off before acceptable performance is reached, leading to a long delay until the expert system is ready for deployment. Additionally, we cannot expect to reach high performance even after a long time has passed, since CBR is limited to cases similar to previously seen cases. To cover the last few percent of cases, the expert system needs to learn about rare situations. Since a rare case, by definition, probably appears for the first time late after the expert system is deployed, we cannot expect the CBR method to handle it well.

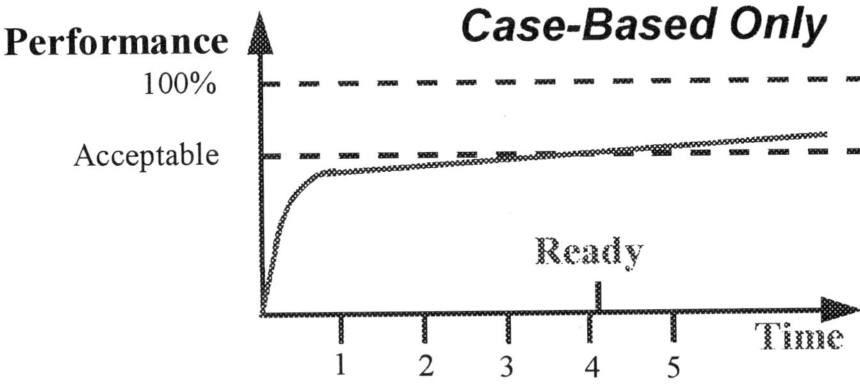

Figure 8: Performance growth for an Only-CBR Diagnostic Expert System

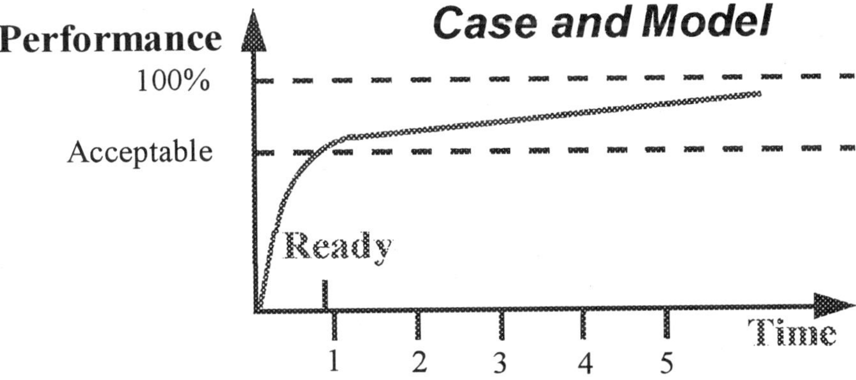

Figure 9: Performance growth for combined MBR/CBR Diagnostic Expert System

Figure 9 shows the typical performance over time of a combined MBR/CBR expert system. Initially, we get the rapid performance rise that is characteristic of CBR methods. However, this rise does not level off as it does with O-CBR, because of the Model-Based component that quickly picks up on the raw case data and converts it to a deeper model. This leads to a very short time to deployment. Moreover, we also get the high eventual performance that is characteristic of MBR methods.

17. CASE STUDIES WITH CBR AND WITH COMBINED MBR/CBR

The principles described here have been repeatedly field proven. For example, Eshkol and Bahir (1997) report a case study of a diagnostic expert system which has been deployed and is now fully operational, supporting repair and maintenance in a large military organization. The expert system was implemented using combined MBR and CBR and was delivered in 40%

less time than originally projected, using model-based knowledge entry for initial population of the knowledge base, followed by a period of case-based learning to improve the performance and add real-world information that was not known, or not expressed by, the expert service technicians. In other expert systems implemented using this approach, users derived cases from existing fault charts and used them to refine the initial model.

In a recent experiment with an O-CBR system, five field engineers started by collecting 70 real historical cases. When new cases (Test Set #1) were submitted for diagnosis, the O-CBR tool came up with too many partial matches yielding a very crude diagnosis; i.e. a large number of possible diagnoses for each new case. To refine the diagnosis, they simulated about 200 more cases in order to differentiate between competing diagnostic alternatives for the cases of Test Set #1. With the expanded 280 training cases, they now tried Test Set #2. The results were still very poor, below 50% correct diagnosis. Considering the effort invested of 4-5 months with 3-6 people, this is very disappointing. The bottom line validates the conclusion reached above: is that with too few cases, the performance would be way below the minimal acceptable level. Unless you have a clear cut case for a given fault, this fault cannot be clearly isolated. On the other hand, collecting a large number of cases is very slow and highly costly.

It is instructive to mention an anecdote from the experience of one TechMate user: In the initial stages developing a knowledge base, a service engineer called from an on-site visit with a problem that he wasn't managing to solve, although he had already replaced several parts. The equipment in question was a large, complex and expensive semiconductor manufacturing machine, and the failure was holding up production. When the engineers at the service headquarters could not remember a similar case either, someone proposed to ask the expert system. The expert system came up with dozens of alternatives, but all except one were already ruled out by the tests performed by the service engineer on-site. The one remaining option was indeed the problem - a defective micro-switch that was erroneously reporting a door as open and therefore triggering a safety interlock circuit and stopping the machine. TechMate's knowledge entry methodology asked the expert engineer to highlight the different circuits involved in correct functionality, and identify the components in each circuit. This method naturally brought to the surface a fact about the micro-switch and the interlock circuit—a fact that nobody remembered, since it was never before involved in any failure.

18. GENERATING FEEDBACK FROM GATHERED CASES INTO ENGINEERING MODELS

A major feature of the method presented here is that the case-base and the model-base interact (Figure 6): Specifically, the cases are not only used for raw access, but also to discover model knowledge, such as new malfunctions. The automatic learning enters the new knowledge features into the model, and refines and cross-correlates other parts of the model to accommodate these additions. The end-result is a model that not only has better-performing diagnostics, but can also be compared side-by-side with the original engineering knowledge derived from the design data (e.g. block diagrams, signal flows). Such comparison, expressed in the same language as the original design data, is very useful for feedback to the engineers responsible for design, testing and reliability analysis. Compare this to the difficulty of extracting meaningful design-model or reliability-model feedback from collections of case reports. The process of using case reports to update the diagram is shown in Figure 8.

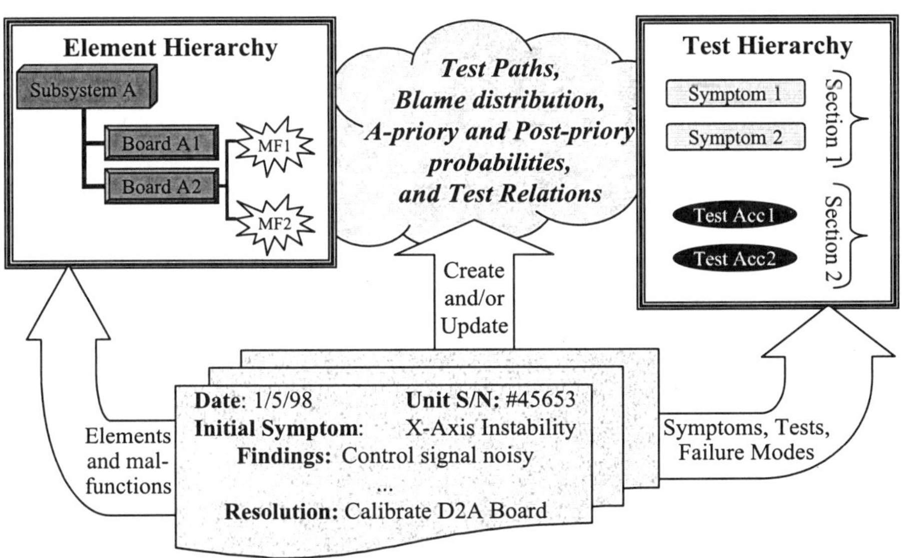

Figure 8: Performance growth for an Only-MBR Diagnostic Expert System

19.　SUMMARY

Many of the CBR deficiencies described in this chapter can be corrected using more advanced methods. Indeed, we see CBR as an integral component of any diagnostic support tool. However, any method that ignores available knowledge and attempts to reverse-engineer it from evidential data is vulnerable to the same inefficiencies and limitations presented here.

To teach an O-CBR expert system, the bare minimum required is one case for each malfunction of each replaceable element. In addition to the large number of required cases, real-life considerations require applying a structure to these cases to overcome the weaknesses caused by a lack of more powerful modeling tools. When the unit under test undergoes any kind of technical modification, the engineer maintaining the case base must again go over the case-base to manually discover which cases need to be added, deleted or modified.

Against these weaknesses of O-CBR, it has always been perceived that MBR has a difficult and lengthy start-up because of the need to explicitly acquire and state the full model. This perception does not always lead to the right conclusions when MBR is compared with CBR. However, our research has shown that a combined MBR-CBR method can overcome the real causes that have led to such perceptions.

The extra effort of creating a large number of cases, painstakingly checking for compliance with the imposed structure, and maintaining the case base after UUT modifications, are all eliminated by combining Case-Based methods with Model-Based methods. There is a wide spectrum of choice: You may start with putting a lot of knowledge into your model, or you may start with almost no knowledge and let accumulated cases perform the dual functions of a basis for Case-Based Reasoning and of a basis for deducing and refining the model. In this chapter, we show that some knowledge is almost always available, and that starting with such knowledge drastically reduces the number of cases required for satisfactory diagnostic performance, as well as alleviating the problems associated with pruning, structuring and updating the case base.

As a final note, we should bear in mind the other advantages of using a deeper kind of knowledge: As employees change jobs inside and outside the organization, precious knowledge about the inner workings of the equipment is lost. Some knowledge never reaches the service group, but stays within the design groups who have long since moved to other projects and have no time to help the support engineers. Any methods of eliciting knowledge, preserving it and moving it along the knowledge chain should yield important benefits for the organization's efficiency over the long run.

20.   REFERENCES

Bearse T. M., "Deriving a Diagnostic Inference Model from a Test Strategy", *Proceedings of the 1998 IEEE International Workshop on System Test and Diagnosis*.

Ben-Bassat M., Beniaminy I., Joseph D., "Different Approaches to Fault Isolation Support Software", *Proceedings of the 1998 IEEE International Workshop on System Test and Diagnosis*.

Ben-Bassat M., Beniaminy I., Eshel M., Sella M., "Practicing Diagnostic Expert Systems for Equipment Maintenance: The AITEST Experience", *Proceedings of the TEST 1992 Conference*, November 1992, Brighton, England.

Ben-Bassat M., Ben-Arie D., Beniaminy I., Cheifetz J. and Klinger M., "AITEST - A Real Life Expert System for Electronic Troubleshooting (A Description and a Case Study)," *Proceedings IEEE Conference on AI Applications*, San Diego (1988).

Ben-Bassat M., "Expert Systems for Clinical Diagnosis," in *Approximate Reasoning in Expert Systems*, M. M. Gupta, A. Kandel, W. Bandler, Y. B. Kiszka (eds.), North Holland, 1985, pp. 671-687.

Bobrow D. G., *Qualitative Reasoning About Physical Systems*, MIT Press, Cambridge, MA, 1985.

Carbonell J. G. (Ed.), *Machine Learning: Paradigms and Methods*, MIT Press, Cambridge, MA, 1989, 1992 (2nd printing).

Dasarathy B. (Ed.), *Nearest Neighbor Pattern Classification*, IEEE Computer Society Press, Los Alamitos, CA 1990.

Eshkol A. and Bahir E., "Deploying Intelligent Support Systems: Practical Application and Lessons Learned," *Proceedings of SOLE: 13th International Logistics Congress*, Jerusalem, 1997.

Fenton W., Webber J., "RF/Microwave Test Simulator," *AUTOTESTCON 1994*.

Lenat D. B. and Feigenbaum E. A., "On the Thresholds of Knowledge," *IJCAI-10*, pp. 1173-1182, Milan, August 1987.

Maguire, Richard J. and Sheppard, JW., "Application Scenarios for AI-ESTATE Services," *AUTOTESTCON 1996 Proceedings*, 1996

Muggleton S. and Feng C., "Efficient Induction of Logic Programs," *Proceedings of the First Conference on Algorithmic Learning Theory*, Tokyo, Japan, Ohmsha, 1990.

Orlidge, Leslie A, "An Overview of IEEE P1232 AI-ESTATE: The Standard for Intelligent Reasoning Based Systems Test and Diagnosis Arrives," *AUTOTESTCON 1996 Proceedings*, 1996

Pazzani M. J. and Brunk C., "Finding Accurate Frontiers: A Knowledge-Intensive Approach to Relational Learning," *The National Conference on Artificial Intelligence*, pp. 328-334. Washington, D.C: AAAI Press.

PC AI Editors: *ai@work, Personal Computer Artificial Intelligence Journal*, November/December 1997, page 16.

PC AI Editors: "New Technology in a Large Customer Call Center," *Personal Computer Artificial Intelligence Journal*, November/December 1997, pp. 24-47.

Quinlan J., "Learning Logical Definitions From Relations," *Machine Learning*, 5 (3), 239—266, 1990.

Sheppard J. W., "Inducing Information Flow Models from Case Data," *Proceedings of the 1998 IEEE International Workshop on System Test and Diagnosis*.

Simpson W. R., Sheppard J. W., *System Test and Diagnosis*, Norwell, MA: Kluwer Academic Publishers, 1994.

Chapter 11

Enhanced Sequential Diagnosis

Anton Biasizzo, Alenka Žužek, Frank Novak
Jozef Stefan Institute

Keywords: System diagnosis, test sequencing, AND/OR graphs, graph search, asymmetrical tests, multivalued tests.

Abstract: The Sequential Diagnosis Tool for the generation of solutions of the test sequencing problem is presented. In contrast to the conventional approach based on the symmetrical and binary tests, the advanced features of the tool are the inclusion of asymmetrical tests and tests with multiple value outcomes. The approach is illustrated by two case studies: diagnosis of boundary scan interconnection test and system level test. The tool also represents the basis of a system diagnosis software package for system maintenance and repair.

1.   INTRODUCTION

Growing demands on the system performance and recent advances in the very large-scale integration technology have resulted in an increasing complexity of electronic systems. While such complexity is needed for the system performance, the problem of maintaining and repairing these systems becomes more and more difficult. The primary goal of system maintenance is to keep the system able to perform designated tasks. If the system fails, the system maintenance has to diagnose and repair the detected failures as rapidly as possible to return the system to the correct operation.

The *fault-free* system operation and the system operation in the presence of a fault can be presented by different *diagnostic states*. The goal of a diagnostic procedure is to identify the actual diagnostic state. The diagnostic procedure isolates the actual diagnostic state by evaluating tests which provide some information on the diagnostic states. In principle, any measurement, signal, or other observable event can be viewed as a system test.

Determining the sequence of steps required to reach a diagnostic conclusion at minimum cost is known as the test sequencing problem. This subject has been studied for more than three decades, and a number of approaches for the solution of the problem have been proposed (Brulé, Johnson, and Kletsky, 1960). NP hardness of the problem has been proved in (Hyafil and Rivest, 1976).

In recent years, the approach based on integrating concepts from information theory and heuristic AND/OR graph search methods has given promising results (Mahanti and Bagchi, 1985; Pattitpati and Alexandridis, 1990). In (Pattipati and Alexandridis, 1990), the algorithm AO* with the Huffman code length-based heuristic search strategy is presented, yielding optimal solutions to problems that were intractable with the existing approaches.

Traditional test sequencing problem is defined for symmetrical binary tests. In the following we first describe the extension of the problem including also multivalued and asymmetrical tests. Next, typical search algorithms for the generation of diagnostic test sequence are briefly described. Finally, illustrative examples employing multivalued and asymmetrical tests are given.

2. TEST SEQUENCING PROBLEM

2.1 System Tests

Any action that reveals a faulty behavior of the system is considered as a test. It consists of two basic operations:

- application of stimuli,
- measurement of the system response.

Tests are evaluated in accordance with employed fault model, and the quality of a test is usually expressed in terms of fault coverage. Notice however that for the purpose of system diagnosis it is not sufficient to achieve high test fault coverage. Besides, a set of tests $T=\{t_1, t_2,..., t_N\}$, which can distinguish individual diagnostic state is prerequisite for the fault isolation.

2.2 Diagnostic States

The operation of the system under test is presented by diagnostic states. Each modeled fault f_i in the system under test that changes the behavior of the system is associated with the corresponding diagnostic state s_i. The *fault-free* operation of the system is represented by a special (*fault-free*) diagnostic state s_0. All different operations of the system with modeled faults are presented by a set of the diagnostic states $S=\{s_0, s_1, ..., s_M\}$, where M is the number of modeled faults. If the system under test is known to be faulty, than the *fault-free* diagnostic state s_0 can be omitted from the set of diagnostic states S.

All faults in the system are not equally frequent, thus an *a-priori* probability is assigned to each system fault. The diagnostic state probabilities are gathered in the *a-priori* probability vector $p=[p_0, p_1,..., p_M]$ representing the probability distribution of the system faults (i.e., diagnostic states).

Fault Dictionary—The execution of a test provides some information on the diagnostic state. Basically, two test outcomes are defined: *pass* and *fail*. For a given fault (i.e., diagnostic state) the test is said to *pass*, when it does not detect the fault and it *fails*, when it detects the fault. Such a binary test t_j divides the set of possible diagnostic states into two sets: $S^{(t_j, 0)}$ and $S^{(t_j, 1)}$, where $S^{(t_j, 0)}$ denotes the set of the remaining diagnostic state candidates, when the test t_j passes and $S^{(t_j, 1)}$ denotes the set of the remaining diagnostic states candidates, when the test fails.

With regard to the properties of these two sets, the following two cases may occur:

- $S^{(t_j, 0)}$ and $S^{(t_j, 1)}$ are disjoint. Test t_j with disjoint sets is called a *symmetrical* test,
- $S^{(t_j, 0)}$ and $S^{(t_j, 1)}$ have nonempty intersection. In this case, there exists at least one diagnostic state s_i, which remains on the candidate list, regardless of the test outcome. For s_i the outcome of t_j is arbitrary (t_j may pass or fail). Test t_j is called an *asymmetrical* test.

The traditional test sequencing problem formulation (Pattipati and Alexandridis, 1990; Pattipati and Dontamsetty, 1992) relies on the symmetrical binary tests. The diagnostic capability of a test t_j is captured by a binary vector d_j, where $d_{ji} = 1$ denotes that test t_j fails if the system is in the diagnostic state s_i. The fault dictionary for such a system is a binary test matrix $D = [d_j]$. The fault dictionary corresponds to the logical closure of the inference matrix (Simpson and Sheppard, 1994).

Inference Engine—Let us assume that p is an *a-priori* probability distribution of a set S of the diagnostic states. When a test t_j is executed the set S is divided into sets $S^{(t_j, 0)}$ and $S^{(t_j, 1)}$. In both sets the probability distribution has changed since some diagnostic states were eliminated. The *a-posteriori* probability distributions of the resulting sets $S^{(t_j, r)}$, where r denotes test outcome (0 for *fail* and 1 for *pass*), are determined by Bayesian inference.

The probability that the test t_j fails is

$$p(t_j = 1) \;=\; \sum_{i=0}^{M} d_{ji} \cdot p(s_i)$$

and the probability distribution of a new set of the diagnostic states $\boldsymbol{S}^{(t_j,\,1)}$, under the assumption that the test t_j has failed is

$$p(s_i \mid t_j = 1) \;=\; \frac{p(s_i)}{p(t_j = 1)}.$$

The probability distribution of the set $\boldsymbol{S}^{(t_j,\,0)}$ is determined in a similar way.

Example 1—An example of a system with symmetrical binary tests is presented in Table 1. The logical inference closure is given in Table 1.a and the corresponding fault dictionary is given in Table 1.b.

Table 1: Logical inference closure and a fault dictionary for Example 1

	s_0	s_1	s_2	s_3	s_4
t_1	x	x		x	
t_2		x	x		x
t_3			x	x	
t_4	x	x	x		

	s_0	s_1	s_2	s_3	s_4
t_1	1	1	0	1	0
t_2	0	1	1	0	1
t_3	0	0	1	1	0
T_4	1	1	1	0	0
$p(s_i)$	0.5	0.2	0.1	0.1	0.1

a)   b)

In Figure 1, the inferences drawn from the fault dictionary, when the test t_2 is executed, are depicted. If the test t_2 fails, the set of the diagnostic state candidates shrinks to $\{s_1,\, s_2,\, s_4\}$. When the test t_2 passes, diagnostic states s_0 and s_3 remain on the candidate list.

Multivalued Tests—In reality, a test may have an arbitrary number of possible outcomes. We denote this feature as *multivalued* test. Let us define a set of test outcomes $\boldsymbol{R}_j = \{r_1,\, r_2,\, ...,\, r_L\}$ of a test t_j. As in the case of binary tests, the fault dictionary is composed of the binary vectors, but now we have to assign L_j vectors $\boldsymbol{d}_j^{(rl)}$, $l=1,\, 2,\, ...,\, L_j$ to the test t_j[1]. It can be shown that L_j-

---

[1] If the tests are known to be symmetrical, it would be easier to capture the diagnostic capabilities of a test t_j by an integer vector $\boldsymbol{d}_j$, where d_{ji} denotes the outcome of the test t_j

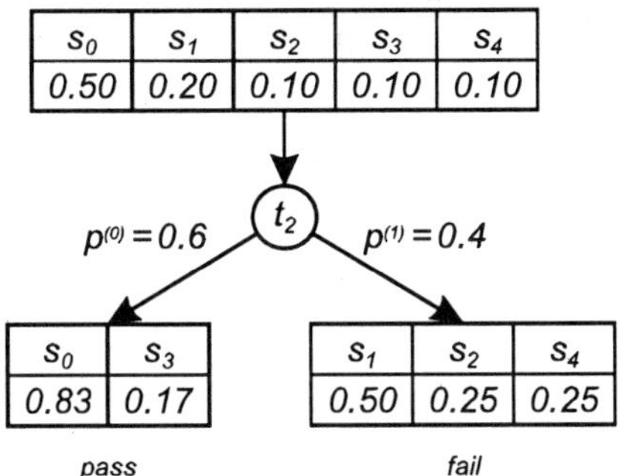

Figure 1: Inferences drawn while executing test t_2

1 binary vectors are sufficient, but for simplicity reasons let us assume that a binary vector is assigned to each test outcome. In the case of a multivalued test, $d_{ji}^{(rl)} = 1$ denotes that the outcome of the test t_j is r_l when the system is in the diagnostic state s_i.

Again, the fault dictionary of the multivalued tests corresponds to the logical closure of the inference matrix (Simpson and Sheppard, 1994), where multivalued tests are presented as a series of binary tests that are tied together and executed simultaneously.

Asymmetrical Tests—As mentioned before, a test t_j may not divide the set of diagnostic states into disjoint sets. We denote it as *asymmetrical* test. Let us assume that a diagnostic state s_i is a member of the sets $\boldsymbol{S}^{(t_j, r_k)}$, $k = 1, 2, ...,$ K_i, where K_i denotes the number of sets which contain the diagnostic state s_i. In this case the *a-priori* probability of the state s_i is distributed among the sets $\boldsymbol{S}^{(t_j, r_k)}$. In order to calculate the *a-posteriori* probabilities of the diagnostic state s_i as well as for calculating the probability that the outcome of test t_j will be r_k, the partitioning $p(s_i, t_j{=}r_k)$ of the *a-priori* probability of the state s_i has to be determined. By applying Bayesian rule

$$p(s_i, t_j = r_k) = p(t_j = r_k \mid s_i) \cdot p(s_i)$$

it can be calculated from the conditional probabilities $p(t_j{=}r_k \mid s_i)$ of the test outcomes r_k if the system is in the diagnostic state s_i. These conditional probabilities form a new fault dictionary. Thus the fault dictionary for the asymmetrical multivalued tests is the test matrix composed of real vectors $\boldsymbol{D}$ $= [\,\boldsymbol{d}_j^{(rl)}\,]$, where $d_{ji}^{(rl)} = p(t_j = r_l \mid s_i)$, $l = 1, 2, ..., L_j$.

---

if the system is in the diagnostic state s_i. However, if the test t_j is asymmetrical, this approach is not applicable because more than one test outcome is possible.

Inference Engine for Asymmetrical Tests—The inference engine for the asymmetrical multivalued tests is similar to the inference engine for the symmetrical binary tests.

The probability that the outcome of a test t_j is r_k is

$$p(t_j = r_k) \;=\; \sum_{i=0}^{M} d_{ji}^{(r_k)} \cdot p(s_i),$$

and a probability distribution of a set of the diagnostic states $S^{(t_j,\, r_k)}$, under the assumption that the outcome of the test t_j is r_k, is

$$p(s_i \,|\, t_j = r_k) \;=\; \frac{d_{ji}^{(r_k)} \cdot p(s_i)}{p(t_j = r_k)}.$$

The fault dictionary D for the asymmetrical tests provides more information than the corresponding logical closure of the inference model. In the inference model, the set of the remaining diagnostic state candidates is given for all test outcomes. This is sufficient to determine the new set of the remaining state candidates, but in the case of an asymmetrical test the probability distribution of the remaining diagnostic states cannot be computed.

Example 2—Let us consider an example of the asymmetrical ternary test t with logical closure given in Table 2.

Table 2: Logical closure for an asymmetrical ternary test t

	s_0	s_1	s_2	s_3	s_4
r_1	X	X			
r_2		X		X	
r_3			X		X
$p(s_i)$	0.50	0.20	0.10	0.10	0.10

In Figure 2, the inferences drawn from the execution of this test are depicted. Since the diagnostic state s_1 remains in the set $S^{(r_1)}$ and $S^{(r_2)}$ the probability of the test outcomes r_1 and r_2 cannot be determined and thus the *a-posteriori* probabilities of the corresponding sets cannot be calculated. The probability that the outcome of the test is r_1 lies in the range $(0.50, 0.70)$, and depends on how frequent the diagnostic state s_1 appears in $S^{(r_1)}$ and $S^{(r_2)}$.

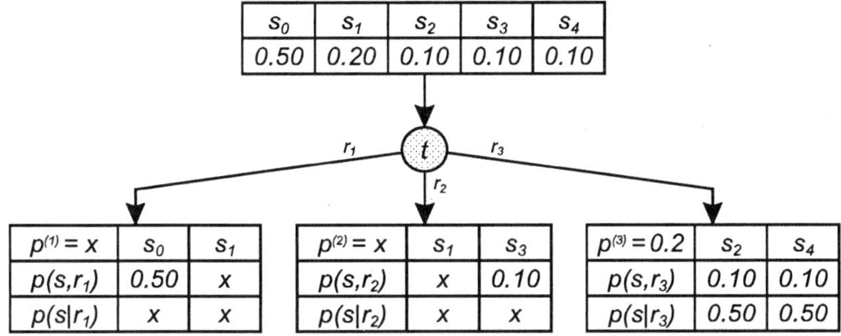

Figure 2: Inferences drawn from the execution of the asymmetrical ternary test t

Suppose that if the system is in the diagnostic state s_1 then both test outcomes r_1 and r_2 are equally probable:

$$p(t_j = r_1 | s_1) \ = \ p(t_j = r_2 | s_1) \ = \ 0.5$$

The corresponding part of the fault dictionary is given in Table 3.

Table 3: Fault dictionary for the asymmetrical ternary test t

	s_0	s_1	s_2	s_3	s_4
r_1	1	0.5	0	0	0
r_2	0	0.5	0	1	0
r_3	0	0	1	0	1
$p(s_i)$	0.50	0.20	0.10	0.10	0.10

With information given in the fault dictionary all *a-posteriori* probabilities are determined. The drawn inferences with the corresponding probabilities are depicted in Figure 3.

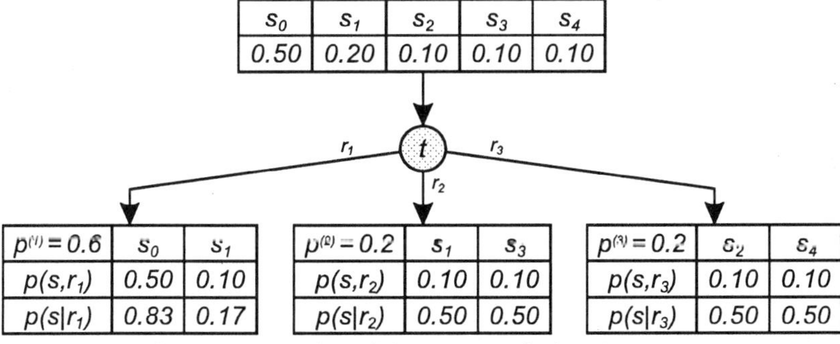

Figure 3: Execution of the asymmetrical ternary test t

2.3   Diagnostic Procedure

A diagnostic procedure is a sequence of tests that is able to isolate any diagnostic state. It can be presented as a decision tree. An example of a diagnostic procedure for Example 1 is shown in Figure 4.

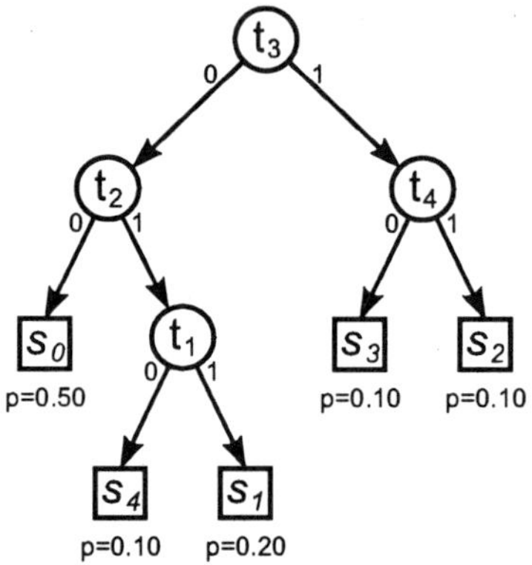

Figure 4: Diagnostic procedure for Example 1

2.4   Cost Function

For a given system there exist a vast number of possible diagnostic procedures. One of the criteria for the selection of the diagnostic procedure can be the average length of the diagnostic procedure

$$\bar{l} = \sum_{i=0}^{M} l_{si}\, p_i,$$

where l_{si} is a number of steps (i.e., tests) needed for the isolation of a diagnostic state s_i. The average length of the diagnostic procedure for Example 1 shown in Figure 4 is $\bar{l} = 2.3$.

The average length of a diagnostic procedure is a good cost function only when all tests are equal. In reality, this is rarely the case. More often some tests are preferred than others, some require specific test equipment while others require well trained operator for the test execution. To cope with different aspects of tests, a cost of a test execution is added to the description of tests.

The test costs can be measured in terms of execution time, operation of the system during test execution (*on-line/off-line* testing), required test equipment, the requirement of a human operator or any other economic factors. When determining the test costs, diagnostic strategic goals play a major role. For example, if the real-time operation of the system is essential, *on-line* tests are preferable then *off-line*, even if they require more sophisticated test equipment and are more time consuming. In the production testing the time-to-market aspect is crucial therefore fast tests which usually require special test equipment are used. It is obvious, that there is no strict rule how to determine the test costs.

Test costs are gathered into a vector $c = [c_1, c_2, ..., c_N]$. Then, the cost function of the diagnostic procedure is defined as

$$J = \sum_{j=1}^{N} p(t_j) \cdot c_j,$$

where the $p(t_j)$ denotes the probability that test t_j is executed during the diagnostic procedure, and c_j denotes the cost of test t_j The probability $p(t_j)$ equals the sum of the ancestor diagnostic state probabilities or the sum of the *a-priori* diagnostic state probabilities for the current set of diagnostic states S

$$p(t_j) = \sum_{si \in S} p(s_i).$$

3. SEARCH ALGORITHMS

The test sequencing problem can be described by a binary AND/OR graph (Nilsson, 1982) where the OR node presents the choice of different tests and the AND node denotes the test partitioning of the set of candidate diagnostic states. The leaves of the AND/OR graph correspond to the individual failure state (i.e. fault isolation).

The AND/OR graph search algorithms explore the AND/OR graph minimizing the given criterion function J. They can be categorized into two major groups: *one-step look-ahead* search algorithms and *global* search algorithms. The one-step look-ahead search algorithms are fast but they do not provide optimal solution. Moreover, it has been proven that the resulting diagnostic procedure might be very costly compared to the optimal diagnostic procedure (Garey and Graham, 1974). On the other hand, the global search algorithms give an optimal solution but their time and space requirements might be excessive. It has been proven that finding the optimal diagnostic procedure is an NP complete problem (Hyafil and Rivest, 1976), thus there is no known algorithm which could solve it in a polynomial time.

3.1   One-Step Look-Ahead Search Algorithms

One-step look-ahead search algorithms provide a trade-off between optimality and computational complexity. They perform a local step-by-step optimization. Local optimization is performed by selecting a test t_j from the set of applicable tests T, which maximizes (or minimizes) some criterion function

$$t_j = arg\ max\{F(S, t_j, c_j)\}$$

In the following, two such algorithms will be presented: the *information algorithm* (Kletsky, 1960) and the *separation algorithm*.

3.1.1   Information Algorithm

The execution of a test t_j can be viewed as an information source about the diagnostic state. Before the test t_j is executed, the system can be in any diagnostic state from the set S, and we have maximum ambiguity. When the test t_j is executed it provides some information on the actual diagnostic state by dividing the set S into subsets $S^{(rk)}$. The ambiguity decreases by the execution of the test.

The information gain defined by the information entropy offers a natural choice for the criterion function since the entropy can be interpreted as a measure of the ambiguity.

$$f_i(S, t_j, c_j) \ = \ \frac{-\sum_{k=1}^{K_j} p(S^{(rk)}) \cdot \log\big(p(S^{(rk)})\big)}{c_j}.$$

3.1.2   Separation Algorithm

In the information algorithm the calculation of the logarithm is the most time consuming operation. Therefore, we try to simplify the criterion function. To mimic the entropy, the product of the probabilities of the subsets $S^{(rk)}$ is chosen. This function has the same bound values as information gain function: $(\prod_{k=1}^{K_j} p(S^{(rk)}) = 0$ when $p(S^{(rk)}) = 1$ for some r_k). It has also the same extreme point (maximum) as the information gain function. Thus the simplified criterion function is

$$f_P(\boldsymbol{S}, t_j, c_j) = \frac{\displaystyle\prod_{k=1}^{K_j} p(\boldsymbol{S}^{(r_k)})}{c_j}.$$

Example 3—The fault dictionary for Example 3 with diagnostic state probabilities and test costs is given in Table 4. We applied the information algorithm to this example. In this case, the test costs were not taken into account, and we set the costs of the test execution to 1. The computed diagnostic procedure is depicted in Figure 5.

Table 4: Fault dictionary and state probabilities for Example 3

	t_1	t_2	t_3	t_4	t_5	$p(s_i)$
s_0	0	0	0	1	0	0.32
s_1	0	0	1	1	0	0.30
s_2	0.8	0	0	1	1	0.16
s_3	1	0	0.5	0	0	0.12
s_4	1	1	1	1	0	0.10

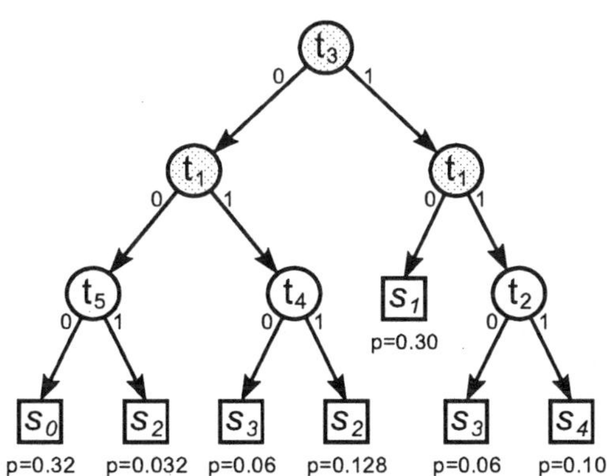

Figure 5: Diagnostic procedure for Example 3

3.2 Global Search Algorithms

Since the construction of the optimal decision tree is an NP complete problem, it is necessary to explore heuristic approaches for guiding the AND/OR graph search (Pattipati and Alexandridis, 1990). These approaches use problem-domain knowledge in the form of a heuristic evaluation function to avoid enumerating the entire set of potential solution trees. The heuristic evaluation function $h(x)$ is an easily computable estimate of the optimal cost-to-go for the node x. The algorithm AO* is one of the well-known AND/OR graph optimization algorithms (Nilsson, 1982). The algorithm AO* is an ordered best-first search algorithm; it expands only the node of the search graph that offers the most promising way of reaching the goal nodes. It has been proven that if the heuristic evaluation function is admissible, the AO* algorithm is guaranteed to find an optimal solution (Mahanti and Bagchi, 1985).

In the case of the symmetrical binary tests, a heuristic evaluation function has been derived (Pattipati and Alexandridis, 1990) by appealing to the analogy between the test sequencing problem and the optimal coding problem. The Huffman coding (Gallager, 1968) serves as the basis of the heuristic evaluation function.

Assuming that the test costs are ordered in ascending order

$$c_1 \leq c_2 \leq c_3 \leq \cdots \leq c_N,$$

the function $\gamma(x)$ is defined as follows

$$\gamma(x) = \sum_{i=1}^{\lfloor x \rfloor} c_i + (x - \lfloor x \rfloor) \cdot c_{\lfloor x \rfloor + 1},$$

where $\lfloor x \rfloor$ denotes the maximal integer number that is not greater than x. The heuristic function is defined as

$$h(S) = p(S) \cdot \gamma\big(w^*(S)\big),$$

where $w^*(S)$ is the average length of the optimal Huffman code for a probability distribution of a set S. The proof of the admissibility of this heuristic function for symmetrical binary tests is given in (Pattipati and Alexandridis, 1990; Pattipati and Dontamsetty, 1992).

The generalization of the test presentation described in Section 2.1.1. arises the question of the validity of the described heuristic function in the general case. It has been proven that the heuristic described above can be employed in the generalized case (Biasizzo, Žužek, and Novak, 1998). Consequently, the conventional algorithms can be employed for the solutions of the generalized test sequencing problem.

The algorithm AO* was applied for the system described in Example 3. Generated diagnostic procedure is depicted in Figure 6.

4. CASE STUDIES

A sequential diagnosis software package including the described algorithms has been implemented. In the following, we report some experimental results in the domains of test sequence generation for system level diagnosis and boundary-scan interconnection tests.

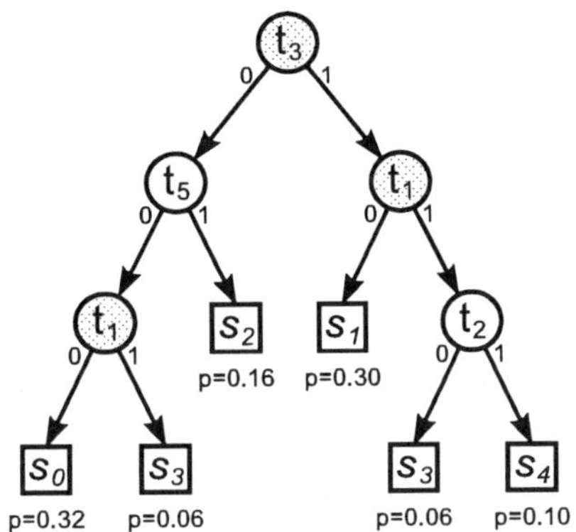

Figure 6: Optimal diagnostic procedure for Example 3

4.1 System-Level Diagnosis

Example 1—The first example of a system-level diagnosis illustrates the conventional test sequencing formulation and is taken from (Simpson and Sheppard, 1994). The system under test is an anti-tank missile launcher and its description is based on the information flow model. Tests in this example are binary symmetrical tests due to the fact that the system has internal test points.

The information flow model comprises two primitive elements, i.e., tests and conclusions. Conclusions in the information flow model correspond to the system faults. Since some conclusions can not be distinguished by a given set of tests, the undistinguished conclusions are gathered in a composite system state. The corresponding associations are given in Table 5.

Table 5: State-conclusions and fault probabilities of anti-tank missile launcher model

State	Conclusions	Probabilities	State	Conclusions	Probabilities
s_1		0.0010	s_9	c_{11}, c_{12}	0.0010
s_2	int_1	0.0010	s_{10}	c_{13}	0.9100
s_3	c_1, c_2	0.0010	s_{11}	c_{14}	0.0100
s_4	c_3	0.0010	s_{12}	c_{15}	0.0100
s_5	c_4	0.0100	s_{13}	c_{16}, c_{17}, c_{18}	0.0025
s_6	c_5	0.0100	s_{14}	c_{19}, inu_1	0.0015
s_7	c_6	0.0100	s_{15}	c_{20}	0.0100
s_8	c_7, c_8, c_9, c_{10}	0.0020	s_{16}	c_{21}	0.0100

This example is described by 16 system states that can be isolated by a given test set. Binary test matrix case is given in Table 6. Test costs are derived by multiplying the required test time by the required skill level. The resulting test cost vector is given in Table 7.

The optimal diagnostic tree for this case is shown in Figure 7. The average cost of the resulting diagnostic tree is $J=4.34$.

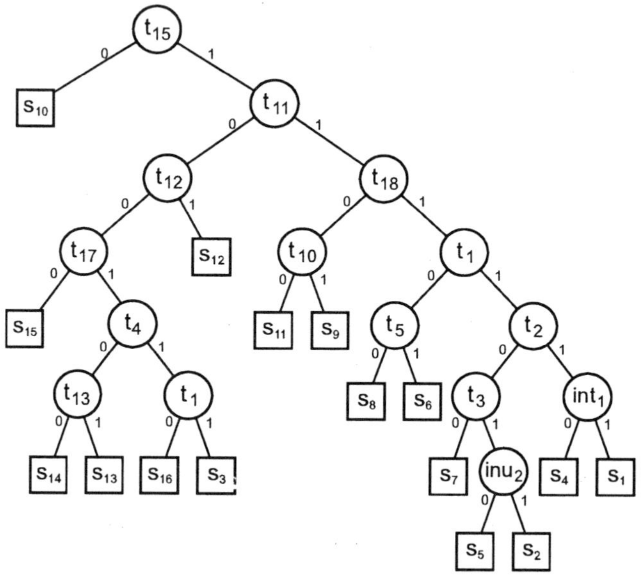

Figure 7: Optimal diagnostic tree for anti-tank missile launcher model

Example 2—The system modeled by strongly interconnected functional blocks and with few internal test points is more difficult to diagnose. Tests are assumed to be applied to primary inputs and the results are observed at primary outputs of the system. Since blocks are mutually connected, the test of a particular block affects other blocks and may detect faults in other blocks as well. This information is captured by asymmetrical tests. As a simple example, let us consider the circuit with three multiplexers and three adders in Figure 8.

Test vectors for each individual module were determined and they were justified to the circuit primary inputs. The results of the applied tests were propagated from the module under test to the circuit primary outputs. The diagnostic capabilities of each developed test were determined by fault simulation.

For illustration, let us consider the test t_{A2} of 2-bit adder A2. First the test vectors which fully test this module were determined with test pattern generation tool. Test vectors are shown in Table 8, where X and Y denote inputs of A2 and Z denotes output.

Table 6: Test matrix of anti-tank missile launcher model

State	t_1	t_2	t_3	t_4	t_5	t_6	t_7	t_8	t_9	t_{10}	t_{11}	t_{12}	t_{13}	t_{14}	t_{15}	t_{16}	t_{17}	t_{18}	int_1	int_2
s_1	1	1	1	1	1	1	1	1	1	1	1	1	1	1	1	1	1	1	1	0
s_2	1	0	1	1	1	1	1	1	1	1	1	1	1	1	1	1	1	1	0	1
s_3	1	0	0	1	0	0	0	0	0	0	0	0	1	1	1	1	1	0	0	0
s_4	1	1	1	1	1	1	1	1	1	1	1	1	1	1	1	1	1	1	0	0
s_5	1	0	1	1	1	1	1	1	1	1	1	1	1	1	1	1	1	1	0	0
s_6	0	0	0	1	1	1	1	1	1	1	1	1	1	1	1	1	1	1	0	0
s_7	1	0	0	1	1	1	1	1	1	1	1	1	1	1	1	1	1	1	0	0
s_8	0	0	0	0	0	1	1	1	1	1	1	1	1	1	1	1	1	1	0	0
s_9	0	0	0	0	0	0	0	0	0	1	1	1	1	1	1	1	1	0	0	0
s_{10}	0	0	0	0	0	0	0	0	0	0	0	0	0	0	0	0	0	0	0	0
s_{11}	0	0	0	0	0	0	0	0	0	0	1	1	1	1	1	1	1	0	0	0
s_{12}	0	0	0	0	0	0	0	0	0	0	0	1	1	1	1	1	1	0	0	0
s_{13}	0	0	0	0	0	0	0	0	0	0	0	0	1	1	1	1	1	0	0	0
s_{14}	0	0	0	0	0	0	0	0	0	0	0	0	0	1	1	1	1	0	0	0
s_{15}	0	0	0	0	0	0	0	0	0	0	0	0	0	0	1	1	0	0	0	0
s_{16}	0	0	0	1	0	0	0	0	0	0	0	0	1	1	1	1	1	0	0	0

Table 7: Test costs of anti-tank missile launcher model

	t_1	t_2	t_3	t_4	t_5	t_6	t_7	t_8	t_9	t_{10}	t_{11}	t_{12}	t_{13}	t_{14}	t_{15}	t_{16}	t_{17}	t_{18}	int_1	int_2
Cost	6.0	6.6	12	6.0	5.2	9.0	5.0	6.0	2.0	3.6	0.7	1.8	3.6	8.0	3.0	6.0	4.5	0.9	2.0	3.0

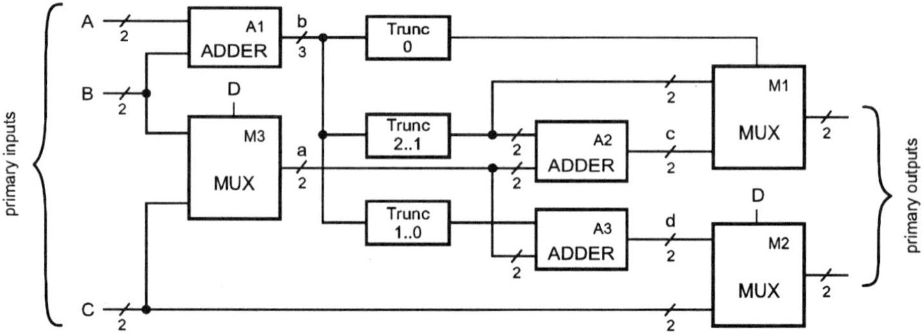

Figure 8: A circuit with three multipliers and two adders

Table 8: Test vectors for adder A2

X	Y	Z
01	10	11
01	01	10
10	00	10
11	01	00
11	10	01
10	01	11
01	11	00
00	00	00
11	11	10

These test vectors were justified to the circuit primary inputs and their responses were propagated to the circuit primary outputs. The obtained primary input test vectors are shown in Table 9. These test vectors form the test t_{A2}.

In Table 10 the stimuli for each module and fault coverage obtained by fault simulation are given. Assuming that *stuck-at* faults within each module are equally probable, the fault coverage can be directly used as a fault dictionary entry.

Table 9: Primary input stimuli and *fault-free* response of the circuit for test t_{A2}

A	B	C	D	Outputs
01	01	10	0	1110
10	00	01	0	1001
10	10	00	0	1000
11	11	01	0	0001
11	11	10	0	0110
01	11	01	0	1101
00	10	11	0	0011
00	00	00	0	0000
11	11	11	0	1011

Table 10: Modul stimuli and fault coverage for test t_{A2}

System parts	M1	M2	M3	A1	A2	A3
Test Vectors	01110	01100	00100	0101	0110	1010
	01100	00010	11010	1000	0101	1001
	10100	10000	00000	1010	1000	0000
	11000	11010	11010	1111	1101	1001
	11010	11100	00100	1111	1110	1010
	10110	11010	01010	0111	1001	0001
	01000	10110	00110	0010	0111	1011
	00000	00000	00000	0000	0000	0000
	11100	11110	00110	1111	1111	1011
Fault coverage	0.8	0	0.6	0.76	1	0

For example, the test t_{A2} fully tests A2 (adder). When this test is applied to the circuit primary inputs, the test stimuli necessary for a complete module A2 test are propagated to the module A2 inputs and the response of this module is propagated to the circuit primary output. On the test stimulus path other modules are excited as well (e.g., M3). But since the test t_{A2} was primarily designed to test the module A2, it does not provide stimuli to completely excite them. Therefore, other tests have to be provided for complete test of the other modules.

Table 11: Test matrix, fault probabilities and test costs for the circuit with three multipliers and two adders

Diagnostic States	Tests										Diagnostic state probabilities
	t_{M1}	t_{M2}	t_{M3}	t_{A1}	t_{A2}	t_{A3}	t_a	t_b	t_c	t_d	
	Test costs c_j										
	1	1	1	2	2	2	10	10	10	10	
fault-free	0	0	0	0	0	0	0	0	0	0	0.43
M1	1	0	0.8	1	0.8	0	0	0	0	0	0.06
M2	0	1	0.9	0.5	0	0.5	0	0	0	0	0.06
M3	0.3	0.5	1	0.5	0.6	0.5	1	0	0.6	0.5	0.06
A1	0.75	0.38	0.49	1	0.76	0.67	0	1	0.76	0.67	0.15
A2	0.61	0	0.45	0.86	1	0	0	0	1	0	0.12
A3	0	0.52	0.61	0.86	0	1	0	0	0	1	0.12

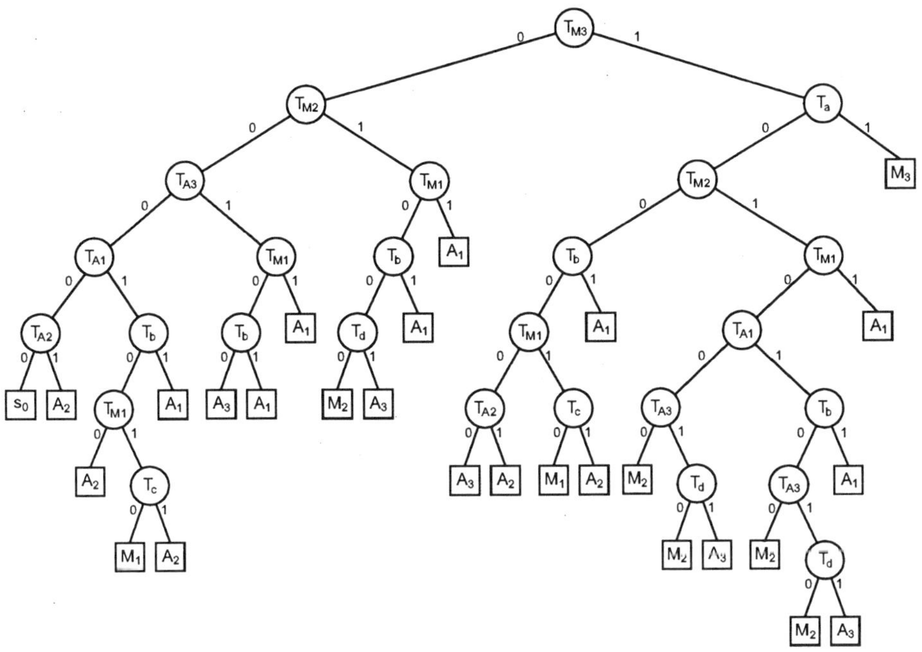

Figure 9: The resulting optimal sequential diagnostic tree

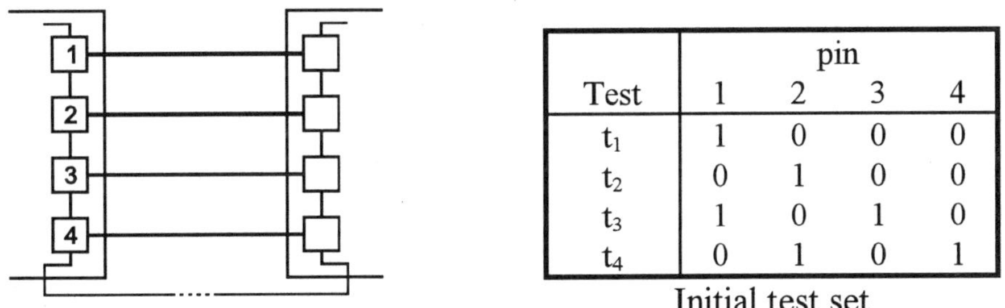

Test	pin			
	1	2	3	4
t_1	1	0	0	0
t_2	0	1	0	0
t_3	1	0	1	0
t_4	0	1	0	1

Initial test set

Figure 10: A set of simple nets

Notice, that with test vectors applied to the primary inputs, it is impossible to distinguish faults of modules connected in series only by observing the primary outputs. Since the objective is to isolate module failures, additional tests (t_a, t_b, t_c, t_d) were introduced. These tests observe internal points of the circuit, and their costs are higher. The diagnostic capabilities of external and internal tests are presented in Table 11. The resulting diagnostic tree for the example of Figure 8 is shown in Figure 9. The average cost of the diagnostic tree is $J = 14.02$.

4.2   Boundary-Scan Interconnection Test

Boundary-Scan architecture is a proven technique for board-level interconnection test. Because of the serial nature of the tests that use boundary-scan, it is important to minimize the test size. The Sequential Diagnosis Tool generates the optimal sequential diagnostic procedure for a given test set.

The algorithm for the generation of the sequential diagnosis was modified to fit the specific behavior of the boundary-scan test technique. In the boundary-scan test technique the next test vector and the previous test results are simultaneously shifted. Thus when the next test vector is applied to the UUT, the result of the previous test is unknown. Resulting delay prolongs the average test sequence length.

Consider a simple example of 4 nets from Figure 10 with all possible bridging faults. This example has 15 possible bridging faults. The initial test set was derived by the *Max-Independence Algorithm* (Jarwala, 1989), but any test set with full diagnostic resolution could be used. The sequential diagnosis software generated an optimal diagnostic tree of average length 2.0, as depicted in Figure 11.

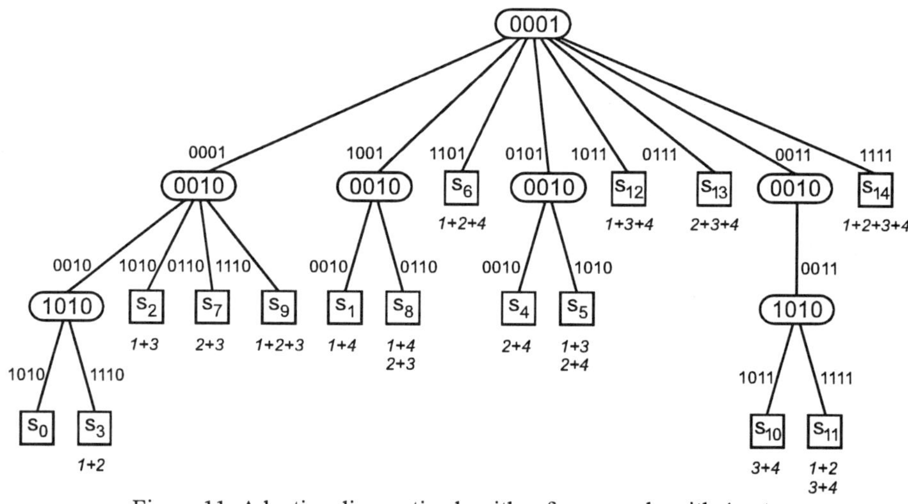

Figure 11: Adaptive diagnostic algorithm for example with 4 nets

Several experiments with different numbers of nets were performed. The number of diagnostic states has increased rapidly since all bridging faults and multiple faults have been assumed. In order to reduce the number of diagnostic states for the bigger examples, only a limited number of faults and shorted nets were considered. For all experiments the initial test sets were derived by the *Max-Independence Algorithm*. Table 12 shows the number of diagnostic states, the achieved average and the maximal cost (length) of diagnostic trees obtained by the sequential diagnosis software for all possible faults considering multiple faults situations. Notice, that the average cost (i.e., the number of actually executed tests) of achieved optimal test sequence is considerably smaller than the initial test size (i.e., the number of tests executed by the conventional approach). The longest test sequence of the computed decision tree is also shorter than the initial test size. Table 13 presents some results on the examples having up to 9 nets with at most triple bridging faults and a limited number of shorted nets.

Table 12: Experimental results for the sequential diagnosis of board-level interconnections with multiple bridging faults

circuit nets	number of states	initial test size	average cost	maximal cost
4 nets	15	4	2.00	3
6 nets	203	6	2.72	4
7 nets	877	7	3.18	5

Another way of reducing the number of diagnostic states (i.e., fault situations) could be to apply structural testing strategy (Salinas, 1996), where only the bridging faults of adjacent nets are considered. In this case, the wiring layout must be specified.

Table 13: Experimental results for the sequential diagnosis of board-level interconnections with most probable bridging faults

circuit nets	shorted nets	number of states	initial test size	average cost	maximal cost
7 nets	4	316	7	3.04	5
8 nets	4	729	8	3.21	6
9 nets	3	1003	9	3.15	6

The sequential diagnosis software was applied to Benchmark deutsch. Benchmark deutsch contains 72 nets and 175 pins. To reduce the number of failure states, the structural testing strategy was applied. All adjacent nets were determined by considering wire intersections. Due to the large number of intersections and adjacencies in the layout, only single faults were considered in our experiment. Resulting optimal diagnostic procedure has the average length $J = 3.63$.

5.   CONCLUSION

In this chapter, we presented the Sequential Diagnosis Tool for the generation of solutions of the test sequencing problem. Apart from the symmetrical test, the diagnostic information provided by the asymmetrical and multivalued tests can be employed in the generation of the diagnostic tree. The obtained solutions surpass the results of the conventional approach where only symmetrical and binary tests are considered. Particular advantage is obtained at the system-level diagnosis where the majority of tests are asymmetrical. Besides, inclusion of multivalued tests in a test procedure further decreases average cost of a diagnostic tree. The described Sequential Diagnosis Tool can be used as a kernel of an expert system for system maintenance and repair or for deriving optimal test sequences of an automatic test equipment system.

6.   REFERENCES

Abramovici, M. and M.A. Breuer, A.D. Friedman, *"Digital Systems Testing and Testable Design"*, IEEE press, 1990.

Biasizzo, A., A. Žužek, and F. Novak, "Sequential Diagnosis with Asymmetrical Tests", To be published in *Computer Journal*, 1998.

Ben-Bassat, M. and M. Sella "Major Evaluation for Real Life Maintenance Expert Systems: The AITEST experience," *Proceeding of Euro Maintenance 92 Conference*, Portugal, 1992.

Brulé, J.D., R.A. Johnson, and E.J. Kletsky, "Diagnosis of equipment failures", *IRE Transaction on Reliability and Quality Control*, Vol. RQC-9, pp. 23-34, April 1960.

Gallager, R.G., *"Information Theory and Reliable Communication"*, Wiley, New York, 1968.

Garey, M.R. and R. L. Graham, "Performance bounds on the splitting algorithm for binary testing", *Acta Informatica*, No. 3, pp. 347-355, 1974.

Hyafil, L. and R. Rivest, "Constructing optimal binary decision trees is NP-complete", *Information Processing Letters*, Vol. 5, No. 1, pp. 15-17, 1976.

Jarwala, N., "A New Framework for Analyzing Test Generation and Diagnosis Algorithms for Wiring Interconnections", *Proceeding of International Test Conference 1989*, pp. 63-70, 1989.

Kletsky, E.J., "An application of the information theory approach to failure diagnosis", *IRE Transaction on Reliability and Quality Control*, Vol. RQC-9, pp. 29-39, December 1960.

Liu, R., *"Testing an Diagnosis of Analog Circuits and Systems"*, New York, Van Nostrand Reinhold, 1991, Chapter 7, pp. 187-216.

Mahanti, A. and A. Bagchi, "AND/OR graph heuristic search methods", *Journal of the ACM*, Vol. 32, No. 1, pp. 28-51, January 1985.

McBean, D. and W.R. Moore, "Testing Interconnects: A Pin Adjacency Approach", *Proceeding of European Test Conference*, pp. 484-490, 1993.

Nilsson, N.J., *"Principles Of Artificial Intelligence"*, Springer-Verlag, 1982, Chapter 3, pp. 99-109.

Pattipati, K.R. and M.G. Alexandridis, "Application of heuristic search and information theory to sequential fault diagnosis", *IEEE Transaction on Systems, Man, and Cybernetics*, Vol. 20, No. 4, pp. 872-887, July/August 1990.

Pattipati, K.R. and M. Dontamsetty, "On a Generalized Test Sequencing Problem", *IEEE Transaction on System, Man and Cybernetics*, Vol. 22, No. 2, pp. 392-396, March/April 1992.

Salinas, J., "A Sweeping Line Approach to Interconnect Testing", *IEEE Transactions on Computers*, Vol.45, No.8, August 1996.

Simpson, W.R. and C.S. Dowling, "WRAPLE: The Weighted Repair Assistance Program Learning Extension", *IEEE Design & Test of Computers*, pp. 66-73, April 1986.

Simpson, W.R. and J.W. Sheppard, *"System Test and Diagnosis"*, Massachusetts, Kluwer Academic Publishers, 1994.

Yeung, R.W., "On Noiseless Diagnosis", *IEEE Transactions on Systems, Man, and Cybernetics*, Vol.24, No.7, July 1994.

Žužek, A., A. Biasizzo, F. Novak, "Towards a general test presentation in the test sequencing problem", *Proceedings of 2nd IEEE International On-Line Testing Workshop*, Biarritz, 1996.